SUGARCANE AS BIOFUEL FEEDSTOCK

Advances Toward a
Sustainable Energy Solution

SUGARCANE AS BIOFUEL FEEDSTOCK

Advances Toward a
Sustainable Energy Solution

Edited by
Barnabas Gikonyo, PhD

Apple Academic Press Inc.	Apple Academic Press Inc.
3333 Mistwell Crescent	9 Spinnaker Way
Oakville, ON L6L 0A2	Waretown, NJ 08758
Canada	USA

©2015 by Apple Academic Press, Inc.

First issued in paperback 2021

Exclusive worldwide distribution by CRC Press, a member of Taylor & Francis Group
No claim to original U.S. Government works

ISBN 13: 978-1-77463-550-6 (pbk)
ISBN 13: 978-1-77188-129-6 (hbk)

Library and Archives Canada Cataloguing in Publication

Sugarcane as biofuel feedstock : advances toward a sustainable energy solution/edited by Barnabas Gikonyo, PhD.

Includes bibliographical references and index.
ISBN 978-1-77188-129-6 (bound)
1. Sugarcane. 2. Energy crops. 3. Biomass energy. I. Gikonyo, Barnabas, editor

SB231.S94 2015	633.6'1	C2015-900728-3

Library of Congress Cataloging-in-Publication Data

Sugarcane as biofuel feedstock: advances toward a sustainable energy solution / Barnabas Gikonyo, PhD, editor.

pages cm
Includes bibliographical references and index.
ISBN 978-1-77188-129-6 (alk. paper)
1. Sugarcane. 2. Energy crops. 3. Biomass energy. I. Gikonyo, Barnabas, editor.

SB231.S835 2015	633.6'1--dc23	2015002643

Apple Academic Press also publishes its books in a variety of electronic formats. Some content that appears in print may not be available in electronic format. For information about Apple Academic Press products, visit our website at **www.appleacademicpress.com** and the CRC Press website at **www.crc-press.com**

ABOUT THE EDITOR

BARNABAS GIKONYO, PhD

Barnabas Gikonyo graduated from Southern Illinois University, Carbon-dale, Illinois (2007), with a PhD in organic and materials chemistry. He currently teaches organic and general chemistry classes at the State University of New York Geneseo, along with corresponding laboratories and the oversight of general chemistry labs. His research interests range from the application of various biocompatible, polymeric materials as "biomaterial bridging surfaces" for the repair of spinal cord injuries, to the use of osteoconductive cements for the repair of critical sized bone defects/fractures. Currently, he is studying the development of alternative, non-food biofuels.

CONTENTS

ACKNOWLEDGMENT AND HOW TO CITE

The editor and publisher thank each of the authors who contributed to this book. The chapters in this book were previously published in various places in various formats. To cite the work contained in this book and to view the individual permissions, please refer to the citation at the beginning of each chapter. Each chapter was read individually and carefully selected by the editor; the result is a book that provides a comprehensive and nuanced perspective on the use of sugarcane bagasse as a biofuel feedstock. The chapters included examine the following topics:

- Chapter 1 provides an overview of renewable energy from sugarcane biomass, offering a good introduction to this book's topic.
- Chapters 2 and 3 focus on Brazil's sugarcane-based biofuel production. With Brazil at the world's forefront in using sugarcane biomass for biofuel feedstock, no discussion of this topic would be complete without research from Brazil.
- Chapter 4 is important to the topic, since microbial technologies offer new opportunities for second-generation biofuels, including biofuel from sugarcane bagasse.
- The authors of chapter 5 offer vital information on how biofuel production can take place side-by-side with food production, using different parts of the same plant (in this case sugarcane bagasse and sugarcane).
- Chapter 6 offers research on using yeast to ferment the pentose and hexose sugars present in sugarcane's lignocellulosic to produce "economic ethanol" that does not jeopardize food production.
- The authors of chapter 7 investigate an economical pretreatment method for lignocellulosic ethanol production.
- In chapter 8, the authors evaluate the production of cellulases from a set of filamentous fungi as a potential source for on-site enzyme production.
- The authors of chapter 9 investigate the conversion of both hexose and pentose sugars in the enzymatic hydrolysates of wet exploded sugarcane bagasse in order to study cell growth and ethanol yields from a particular strain of yeast.
- Chapter 10 offers yet another example of second-generation ingenuity, focusing on the feasibility of a flexible biorefinery that could use sugarcane feedstock to produce either ethanol or electricity.

- The authors of chapter 11 confront the fact that biofuels are not the magic answer to the world's environmental problems, since biofuel production also has impact on the environment. Their study evaluates the environmental impact of residues generated during the consecutive acid-base pretreatment of sugarcane bagasse.
- In chapter 12, we turn to considerations for second- and third-generation biofuels. This first study offers decision-making components to this initiative by evaluating actions for improving the performance of electric energy cogeneration units by burning sugarcane bagasse.
- Chapter 13 provides a detailed techno-economic comparison of scenarios that entail ethanol coproduction with export electricity, against those that produce only electricity using the same systems.

LIST OF CONTRIBUTORS

Dilip K. Adhikari
Biotechnology Conversion Area, CSIR-Indian Institute of Petroleum, Mohkampur, Dehradun, UK, 248005, India

Deepti Agrawal
Biotechnology Conversion Area, CSIR-Indian Institute of Petroleum, Mohkampur, Dehradun, UK, 248005, India

Birgitte K. Ahring
Section for Sustainable Biotechnology, Aalborg University Copenhagen, Copenhagen, Denmark; Center for Bioproducts and Bioenergy, Washington State University, 2710 University Drive, Richland, WA 99354-1671, USA

Mathew C Aneke
Department of Process Engineering, University of Stellenbosch, Cnr Banghoek Road & Joubert Street, Stellenbosch 7600, South Africa

Virgilio Anjos
Material Spectroscopy Laboratory, Department of Physics, Federal University of Juiz de Fora,, 36036-330, Juiz de Fora, MG, Brazil

Felipe F. A. Antunes
Department of Biotechnology, University of São Paulo, School of Engineering of Lorena, Estrada Municipal do Campinho- Caixa,, Postal 116 12.602.810, Lorena/SP, Brazil

Luiza Carvalho Martins Arruda
Chemical Engineering Department, Polytechnic School of the University of Sao Paulo, Av. Prof. Lineu Prestes, 580, Bloco 18—Conjunto das Químicas, 05424-970 São Paulo, Sao Paulo, Brazil

L. C. Basso
University of São Paulo, "Luiz de Queiroz" College of Agriculture, Brazil

T. O. Basso
University of São Paulo, "Luiz de Queiroz" College of Agriculture, Brazil

T. P. Basso
University of São Paulo, "Luiz de Queiroz" College of Agriculture, Brazil

Maria J. V. Bell
Material Spectroscopy Laboratory, Department of Physics, Federal University of Juiz de Fora,, 36036-330, Juiz de Fora, MG, Brazil

Bruno Benoliel
Centro de Biotecnologia Molecular, Departamento de Biologia Celular, Instituto de Ciências Biológicas, Universidade de Brasília, Brasília, DF, Brazil

Oigres D. Bernardinelli
Instituto de Física de São Carlos, Universidade de São Paulo, Caixa Postal 369, São Carlos SP 13560-970, Brazil

Rajib Biswas
Section for Sustainable Biotechnology, Aalborg University Copenhagen, Copenhagen, Denmark; Center for Bioproducts and Bioenergy, Washington State University, 2710 University Drive, Richland, WA 99354-1671, USA

Raj Boopathy
Department of Biological Sciences Nicholls State University Thibodaux, LA 70310, USA.

Marcelo A. Carvalho
Embrapa Cerrados, Genética e Melhoramento de Forrageiras, Br 020, Km 18 – Cx. P. 08223, Planaltina DF 73301-970, Brazil

Anuj K. Chandel
Department of Biotechnology, University of São Paulo, School of Engineering of Lorena, Estrada Municipal do Campinho- Caixa,, Postal 116 12.602.810, Lorena/SP, Brazil

Caliane B. B. Costa
Chemical Engineering Graduate Program, Federal University of São Carlos, PPGEQ/UFSCar Via Washington Luis, km 235, São Carlos, SP, Brazil; Department of Chemical Engineering, Federal University of São Carlos, DEQ/UFSCar Via Washington Luis, km 235, São Carlos, SP, Brazil

Antonio J. G. Cruz
Chemical Engineering Graduate Program, Federal University of São Carlos, PPGEQ/UFSCar Via Washington Luis, km 235, São Carlos, SP, Brazil; Department of Chemical Engineering, Federal University of São Carlos, DEQ/UFSCar Via Washington Luis, km 235, São Carlos, SP, Brazil

José Roberto Coleta Jr.
Chemical Engineering Department, Polytechnic School of the University of Sao Paulo, Av. Prof. Lineu Prestes, 580, Bloco 18—Conjunto das Químicas, 05424-970 São Paulo, Sao Paulo, Brazil

Silvio S. da Silva
Department of Biotechnology, University of São Paulo, School of Engineering of Lorena, Estrada Municipal do Campinho- Caixa,, Postal 116 12.602.810, Lorena/SP, Brazil

Diptarka Dasgupta
Biotechnology Conversion Area, CSIR-Indian Institute of Petroleum, Mohkampur, Dehradun, UK, 248005, India

Eduardo R. deAzevedo
Instituto de Física de São Carlos, Universidade de São Paulo, Caixa Postal 369, São Carlos SP 13560-970, Brazil

J. Deckers
Katholieke Universiteit Leuven, Department of Earth and Environmental Sciences, Celestijnenlaan Leuven, Belgium

Fabrícia Faria
Universidade Federal de Goiás, Brazil

Renato Tonon Filho
Chemical Engineering Graduate Program, Federal University of São Carlos, PPGEQ/UFSCar Via Washington Luis, km 235, São Carlos, SP, Brazil

Felipe F. Furlan
Chemical Engineering Graduate Program, Federal University of São Carlos, PPGEQ/UFSCar Via Washington Luis, km 235, São Carlos, SP, Brazil

C. R. Gallo
University of São Paulo, "Luiz de Queiroz" College of Agriculture, Brazil

Debashish Ghosh
Biotechnology Conversion Area, CSIR-Indian Institute of Petroleum, Mohkampur, Dehradun, UK, 248005, India

Leonardo D. Gomez
CNAP, Department of Biology, University of York, York, Heslington YO10 5DD, UK

Johann F. Görgens
Department of Process Engineering, University of Stellenbosch, Cnr Banghoek Road & Joubert Street, Stellenbosch 7600, South Africa

Raquel L. C. Giordano
Chemical Engineering Graduate Program, Federal University of São Carlos, PPGEQ/UFSCar Via Washington Luis, km 235, São Carlos, SP, Brazil; Department of Chemical Engineering, Federal University of São Carlos, DEQ/UFSCar Via Washington Luis, km 235, São Carlos, SP, Brazil

Roberto C. Giordano
Chemical Engineering Graduate Program, Federal University of São Carlos, PPGEQ/UFSCar Via Washington Luis, km 235, São Carlos, SP, Brazil; Department of Chemical Engineering, Federal University of São Carlos, DEQ/UFSCar Via Washington Luis, km 235, São Carlos, SP, Brazil

João Paulo Macedo Guerra
Chemical Engineering Department, Polytechnic School of the University of Sao Paulo, Av. Prof.
Lineu Prestes, 580, Bloco 18—Conjunto das Químicas, 05424-970 São Paulo, Sao Paulo, Brazil

Moses Isabirye
Faculty of Natural Resources and Environment, Namasagali Campus, Busitema University, Kamuli,
Uganda

Rakesh Kumar Jain
Division of Chemical Recovery, Central Pulp and Paper Research Institute, Biotechnology & Lignin
by-products, Saharanpur, UP, 247001, India

Rashmi Khan
Biotechnology Conversion Area, CSIR-Indian Institute of Petroleum, Mohkampur, Dehradun, UK,
248005, India

M. Kitutu
National Environment Management Authority, Kampala, Uganda

Luiz Kulay
Chemical Engineering Department, Polytechnic School of the University of Sao Paulo, Av. Prof.
Lineu Prestes, 580, Bloco 18—Conjunto das Químicas, 05424-970 São Paulo, Sao Paulo, Brazil

Carlos A. Labate
Laboratório Max Feffer de Genética de Plantas, Departamento de Genética -ESALQ, Universidade
de São Paulo, Caixa Postal 83, Piracicaba SP 13418-900, Brazil; Centro Nacional de Pesquisa em
Energia e Materiais, Laboratório Nacional de Ciência e Tecnologia do Bioetanol (CTBE), Campinas,
SP, Brazil

Marisa A. Lima
Instituto de Física de São Carlos, Universidade de São Paulo, Caixa Postal 369, São Carlos SP 13560-
970, Brazil

Simon J. McQueen-Mason
CNAP, Department of Biology, University of York, York, Heslington YO10 5DD, UK

Lidia Maria Pepe de Moraes
Centro de Biotecnologia Molecular, Departamento de Biologia Celular, Instituto de Ciências Biológicas, Universidade de Brasília, Brasília, DF, Brazil; Laboratório de Biologia Molecular, Departamento
de Biologia Celular, Universidade de Brasília, Brasília, DF 70910-900, Brazil

Daniella Moreira
Universidade Federal de Goiás, Brazil

Ludmylla Noleto
Universidade Federal de Goiás, Brazil

Ivy dos Santos Oliveira
Department of Biotechnology, Engineering School of Lorena, University of São Paulo, Estrada Municipal do Campinho, P.O. Box, Lorena/SP 116 12.602.810, Brazil

Fernando C. Pagnocca
Department of Biochemistry and Microbiology, Institute of Biosciences CEIS/UNESP – Rio, Claro/SP, Brazil

Diwakar Pandey
Division of Chemical Recovery, Central Pulp and Paper Research Institute, Biotechnology & Lignin by-products, Saharanpur, UP, 247001, India

Abdul M. Petersen
Department of Process Engineering, University of Stellenbosch, Cnr Banghoek Road & Joubert Street, Stellenbosch 7600, South Africa

Fabio H. P. B. Pinto
Chemical Engineering Graduate Program, Federal University of São Carlos, PPGEQ/UFSCar Via Washington Luis, km 235, São Carlos, SP, Brazil

J. Poesen
Katholieke Universiteit Leuven, Department of Earth and Environmental Sciences, Celestijnenlaan Leuven, Belgium

Igor Polikarpov
Instituto de Física de São Carlos, Universidade de São Paulo, Caixa Postal 369, São Carlos SP 13560-970, Brazil

Bijeta Prasai
Department of Chemistry, Louisiana State University, Baton Rouge, LA 70803, USA.

D. V. N. Raju
Research and Dev't Section - Agricultural Department Kakira Sugar Limited, Jinja, Uganda

Camila A. Rezende
Instituto de Química, Universidade de Campinas, Caixa Postal 6154, Campinas SP 13083-970, Brazil

Leonarde N. Rodrigues
Material Spectroscopy Laboratory, Department of Physics, Federal University of Juiz de Fora,, 36036-330, Juiz de Fora, MG, Brazil

Carlos A. Rosa
Department of Microbiology, Federal University of Minas Gerais,, Belo Horizonte, MG, Brazil

David Samaha
Department of Biological Sciences Nicholls State University Thibodaux, LA 70310, USA.

Gil Anderi Silva
Chemical Engineering Department, Polytechnic School of the University of Sao Paulo, Av. Prof. Lineu Prestes, 580, Bloco 18—Conjunto das Químicas, 05424-970 São Paulo, Sao Paulo, Brazil

Messias Borges Silva
Department of Chemical Engineering, Engineering School of Lorena, University of São Paulo, Lorena 12.602.810, Brazil

Om V. Singh
Division of Biological and Health Sciences, University of Pittsburgh, 16701, Bradford, PA, USA

Rachael Simister
CNAP, Department of Biology, University of York, York, Heslington YO10 5DD, UK

Clare G. Steele-King
CNAP, Department of Biology, University of York, York, Heslington YO10 5DD, UK

V. Sri Harjati Suhardi
School of Life Sciences and Technology, Institut Teknologi Bandung, Jl. Ganesha 10, Bandung 40132, Indonesia.

Sunil Kumar Suman
Biotechnology Conversion Area, CSIR-Indian Institute of Petroleum, Mohkampur, Dehradun, UK, 248005, India

Fernando Araripe Gonçalves Torres
Centro de Biotecnologia Molecular, Departamento de Biologia Celular, Instituto de Ciências Biológicas, Universidade de Brasília, Brasília, DF, Brazil

Hinrich Uellendahl
Section for Sustainable Biotechnology, Aalborg University Copenhagen, Copenhagen, Denmark

Vasanta Thakur Vadde
Division of Chemical Recovery, Central Pulp and Paper Research Institute, Biotechnology & Lignin by-products, Saharanpur, UP, 247001, India

V. Yemeline
UNEP/GRID-Arendal, Norway

INTRODUCTION

While the world's demand for fossil fuels grows each year, we face ominous rumblings that fossil fuel production is no longer a viable option. We can no longer ignore those rumblings. Diminishing levels of underground crude oil and aging refineries comprise only a small portion of the problem we face. Far more serious is the burden on our planet, and the need do whatever we can to mitigate the effects of climate change. Biofuels offer us one possible, partial answer to the world's dilemma.

However, first-generation biofuels that used corn grains or sugarcane juice as substrates offered alternatives fraught with their own dangers, since they promised to cause price hikes and food shortages around the world. Bioethanol produced from sustainable feedstocks such as lignocellulosics offers a far more useful option. As an alternative to both fossil and food-based fuels, cellulosic ethanol brings environmental sustainability within our reach.

The selection of a biomass substrate and its efficient utilization are critical steps in the process of making this happen.

Sugarcane is one of the most efficient crops in converting sunlight energy to chemical energy for fuel. This volume brings together a collection of vital research on this topic, paving the way for ongoing investigations that will lead us to a more sustainable future.

—*Barnabas Gikonyo. PhD*

In chapter one, the researchers' overall objective was too assess the sustainable production of biofuels and electricity from sugarcane biomass within the context of poverty, food insecurity, and environmental integrity. The authors provide an overview of renewable energy from sugarcane biomass, offering a good introduction to this book's topic, discussing the balance between using land resources for food and feed versus for energy provision. This is a particularly sensitive and important consideration in a nation such as Uganda, where malnutrition and food insecurity are major issues. Growth in sugarcane cultivation has resulted in increased demand for land to produce staple foods for households and has encroached on fragile wetland and forest ecosystems. This chapter aims to demonstrate how sugarcane biomass can sustainably be produced to support fuel and electrical energy demands, while at the same time conserving the environment and ensuring increased household income and food security.

In chapter two, the authors discuss the ways in which the search for promising and renewable sources of carbohydrates for the production of biofuels and other biorenewables has been stimulated by an increase in global energy demand in the face of growing concern over greenhouse gas emissions and fuel security. In particular, they stress, interest has focused on non-food lignocellulosic biomass as a potential source of abundant and sustainable feedstock for biorefineries. They investigate the potential of three Brazilian grasses (*Panicum maximum*, *Pennisetum purpureum* and *Brachiaria brizantha*), as well as bark residues from the harvesting of two commercial Eucalyptus clones (*E. grandis* and *E. grandis x urophylla*) for biofuel production, and compare these to sugarcane bagasse. The effects of hot water, acid, alkaline and sulfite pretreatments (at increasing temperatures) on the chemical composition, morphology and saccharification yields of these different biomass types were evaluated. The authors compared the average yield (per hectare), availability, and general composition of all five biomasses. Compositional analyses indicated a high level of hemicellulose and lignin removal in all grass varieties (including sugarcane bagasse) after acid and alkaline pretreatment with increasing temperatures, while the biomasses pretreated with hot water or sulfite showed little variation from the control. For all biomasses, higher cellulose enrichment resulted from treatment with sodium hydroxide at 130°C. At 180°C, a decrease in cellulose content was observed, which is associated with high amorphous cellulose removal and 5-hydroxymethyl-furaldehyde production. Morphological analysis showed the effects of different pretreatments on the biomass surface, revealing a high production of microfibrillated cellulose on grass surfaces, after treatment with 1% sodium hydroxide at 130°C for 30 minutes. This may explain the higher hydrolysis yields resulting from these pretreatments, since these cellulosic nanoparticles can be easily accessed and cleaved by cellulases. The authors' results show the potential of three Brazilian grasses with high productivity yields as valuable sources of carbohydrates for ethanol production and other biomaterials. Sodium hydroxide at 130°C was found to be the most effective pretreatment for enhanced saccharification yields. It was also efficient in the production of microfibrillated cellulose on grass surfaces, thereby revealing their potential as a source of natural fillers used for bionanocomposites production.

In chapter 3, the authors discuss sugarcane bagasse retreatment, hydrolysis, and fermentation using various approaches. They explain that according to a CTC protocol, the process of manufacturing ethanol from bagasse is divided into the following steps. First, the bagasse is pretreated via steam explosion (with or without a mild acid condition) to increase the enzyme accessibility to the cellulose and promoting the hemicellulose hydrolysis with a pentose stream. The lignin and cellulose solid fraction is subjected to cellulose hydrolysis, generating a hexose-rich stream (mainly composed of glucose, manose and galactose). The final solid residue (lignin and the remaining recalcitrant cellulose) is used for heating and steam generation. The hexose fraction is mixed with 1G cane molasses (as a source of minerals, vitamins and aminoacids) and fermented by regular *Saccharomyces cerevisiae* industrial strains (not genetically modified) using the same fermentation and distillation facilities of the Brazilian ethanol plants. The pentose fraction will be used as substrate for other biotechnological purposes, including ethanol fermentation.

In chapter 4, the authors provide a brief summary of seven cultures they selected (A3, B3, M2, M3, X7, F4 and D2) that produce cellulases and xylanases from the sugarcane bagasse. They chose these according to the halo of hydrolysis diameter to determine their enzymatic activity. The culture A3 proved to be a good producer of xylanase. The culture M3 produced cellulases with FPase and CMCase activity, showing that is good for cellulose hydrolysis. The culture X7 simultaneously produces cellulases and xylanases, which favors the hydrolysis of cellulose and hemicellulose using SCB as substrate.

The authors of chapter 5 analyze ethanol production from sugarcane bagasse pith hydrolysate by thermotolerant yeast *Kluyveromyces sp.*, using response surface methodology. Variables such as substrate concentration, pH, fermentation time, and Na2HPO4 concentration were found to influence ethanol production significantly. In a batch fermentation, optimization of key process variables resulted in maximum ethanol concentration of 17.44 g/L which was 88% of the theoretical with specific productivity of 0.36 g/L/h.

Chandel et al. discuss in chapter 6 the ways in which diminishing supplies of fossil fuels and oil spills provide impetus to explore alternative sources of energy that can be produced from non-food/feed-based substrates. They conclude that due to its abundance, sugarcane bagasse could be a model substrate for second-generation biofuel cellulosic ethanol. However, the efficient bioconversion of sugar bagasse remains a challenge for the commercial production of cellulosic ethanol. The authors hypothesized that oxalic-acid-mediated thermochemical pretreatment (OAFEX) would overcome the native recalcitrance of sugar bagasse by enhancing the cellulase amenability toward the embedded cellulosic microfibrils. In fact, OAFEX treatment revealed the solubilization of hemicellulose releasing sugars (12.56 g/l xylose and 1.85 g/l glucose), leaving cellulignin in an accessible form for enzymatic hydrolysis. The highest hydrolytic efficiency (66.51%) of cellulignin was achieved by enzymatic hydrolysis (Celluclast 1.5 L and Novozym 188). The ultrastructure characterization of bagasse using scanning electron microscopy (SEM), atomic force microscopy (AFM), Raman spectroscopy, Fourier transform–near infrared spectroscopy (FT-NIR), Fourier transform infrared spectroscopy (FTIR), and X-ray diffraction (XRD) revealed structural differences before and after OAFEX treatment with enzymatic hydrolysis. Furthermore, fermentation mediated by C. shehatae UFMG HM52.2 and S. cerevisiae 174 showed fuel ethanol production from detoxified acid (3.2 g/l, yield 0.353 g/g; 0.52 g/l, yield, 0.246 g/g) and enzymatic hydrolysates (4.83 g/l, yield, 0.28 g/g; 6.6 g/l, yield 0.46 g/g). Finally, OAFEX treatment revealed marked hemicellulose degradation, improving the cellulases' ability to access the cellulignin and release fermentable sugars from the pretreated substrate. The ultrastructure of sugar bagasse after OAFEX and enzymatic hydrolysis of cellulignin established thorough insights at the molecular level.

The process of converting lignocellulosic biomass to ethanol involves pretreatment to disrupt the complex of lignin, cellulose, and hemicellulose, freeing cellulose and hemicellulose for enzymatic saccharification and fermentation. Determining optimal pretreatment techniques for fermentation is essential for the success of lignocellulosic energy production process. The authors of chapter 7 evaluate energy cane for lignocellulosic ethanol production. Various pretreatment processes for energy cane variety L 79-1002 (type II) were evaluated, including different concentrations of dilute acid hydrolysis and solid-state fungal pretreatment process using brown rot and white rot fungi. Pretreated biomass was enzymatically saccharified and fermented using a recombinant *Escherichia coli*. The results revealed that all pretreatment processes that were subjected to enzymatic saccharification and fermentation produced ethanol. However, the best result was observed in dilute acid hydrolysis of 3% sulfuric acid. Combination of fungal pretreatment with dilute acid hydrolysis reduced the acid requirement from 3% to 1% and this combined process could be more economical in a large-scale production system.

Brazil is a major producer of agro-industrial residues, such as sugarcane bagasse, which could be used as raw material for microbial production of cellulases as an important strategy for the development of sustainable processes of second generation ethanol production. For this purpose, in chapter 8 Benoliel and his colleagues worked to screen for glycosyl hydrolase activities of fungal strains isolated from the Brazilian Cerrado. Among thirteen isolates, a *Trichoderma harzianum* strain (L04) was identified as a promising candidate for cellulase production when cultured on *in natura* sugarcane bagasse. Strain L04 revealed a well-balanced cellulolytic complex, presenting fast kinetic production of endoglucanases, exoglucanases and β-glucosidases, achieving 4,022, $U.L^{-1}$ (72 h), 1,228 $U.L^{-1}$ (120 h) and 1,968 $U.L^{-1}$ (48 h) as the highest activities, respectively. About 60% glucose yields were obtained from sugarcane bagasse after 18 hours hydrolysis. This new strain represents a potential candidate for on-site enzyme production using sugarcane bagasse as carbon source.

Because sugarcane bagasse is a potential feedstock for cellulosic ethanol production, rich in both glucan and xylan, both C6 and C5 sugars need to be used for conversion into ethanol in order to improve the process economics. During processing of the hydrolysate degradation products such as acetate, 5-hydroxymethylfurfural (HMF) and furfural are formed, which are known to inhibit microbial growth at higher concentrations. In the study found in chapter 9, conversion of both glucose and xylose sugars into ethanol in wet exploded bagasse hydrolysates was investigated without detoxification using *Scheffersomyces (Pichia) stipitis* CBS6054, a native xylose utilizing yeast strain. The sugar utilization ratio and ethanol yield

(Yp/s) ranged from 88-100% and 0.33-0.41±0.02 g/g, respectively, in all the hydrolysates tested. Hydrolysate after wet explosion at 185°C and 6 bar O_2, composed of mixed sugars (glucose and xylose) and inhibitors such as acetate, HMF and furfural at concentrations of 3.2±0.1, 0.4 and 0.5 g/l, respectively, exhibited highest cell growth rate of 0.079 g/l/h and an ethanol yield of 0.39±0.02 g/g sugar converted. *Scheffersomyces stipitis* exhibited prolonged fermentation time on bagasse hydrolysate after wet explosion at 200°C and 6 bar O_2 where the inhibitors concentration was further increased. Nonetheless, ethanol was produced up to 18.7±1.1 g/l resulting in a yield of 0.38±0.02 g/g after 82 h of fermentation.

Although sugarcane is the most efficient crop for production of 1G ethanol and can also be used to produce 2G ethanol, the large-scale manufacture of 2G ethanol is not a consolidated process yet. Thus, a detailed economic analysis, based on consistent simulations of the process, is worthwhile. Moreover, both ethanol and electric energy markets have been extremely volatile in Brazil, which suggests that a flexible biorefinery, able to switch between 2G ethanol and electric energy production, could be an option to absorb fluctuations in relative prices. In chapter 10, simulations of three cases were run using the software EMSO: production of 1G ethanol + electric energy, of 1G + 2G ethanol and a flexible biorefinery. Bagasse for 2G ethanol was pretreated with a weak acid solution, followed by enzymatic hydrolysis, while 50% of sugarcane trash (mostly leaves) was used as surplus fuel. With maximum diversion of bagasse to 2G ethanol (74% of the total), an increase of 25.8% in ethanol production (reaching 115.2 L/tonne of sugarcane) was achieved. An increase of 21.1% in the current ethanol price would be enough to make all three biorefineries economically viable (11.5% for the 1G + 2G dedicated biorefinery). For 2012 prices, the flexible biorefinery presented a lower Internal Rate of Return (IRR) than the 1G + 2G dedicated biorefinery. The impact of electric energy prices (auction and spot market) and of enzyme costs on the IRR was not as significant as it would be expected. The authors conclude that for current market prices in Brazil, not even production of 1G bioethanol is economically feasible. However, the 1G + 2G dedicated biorefinery is closer to feasibility than the conventional 1G + electric energy industrial plant. Besides, the IRR of the 1G + 2G biorefinery is more sensitive with respect to the price of ethanol, and an increase of 11.5% in this value would be enough to achieve feasibility. The ability of the flexible biorefinery to take advantage of seasonal fluctuations does not make up for its higher investment cost, in scenario the authors consider.

In chapter 11, Santos Oliveira and her colleagues acknowledge that the production of biofuels from sugar bagasse can have a negative impact on the environment, due to the use of harsh chemicals during pretreatment. Consecutive sulfuric acid-sodium hydroxide pretreatment of SB is an effective process, however, that eventually ameliorates the accessibility of cellulase towards cellulose for the sugars production. The authors' work evaluates the environmental impact of residues generated during the consecutive acid-base pretreatment of SB. Advanced oxidative process (AOP) was used based on photo-Fenton reaction mechanism (Fenton Reagent/ UV). Experiments were performed in batch mode following factorial design L_9 (Taguchi orthogonal array design of experiments), considering the three operation variables: temperature (°C), pH, Fenton Reagent (Fe^{2+}/ H_2O_2) + ultraviolet. Reduction of total phenolics (TP) and total organic carbon (TOC) were responsive variables. Among the tested conditions, experiment 7 (temperature, 35°C; pH, 2.5; Fenton reagent, 144 ml H_2O_2+153 ml Fe^{2+}; UV, 16W) revealed the maximum reduction in TP (98.65%) and TOC (95.73%). Parameters such as chemical oxygen demand (COD), biochemical oxygen demand (BOD), BOD/COD ratio, color intensity and turbidity also showed a significant change in AOP mediated lignin solution than the native alkaline hydrolysate. The authors conclude that AOP based on Fenton Reagent/UV reaction mechanism showed efficient removal of TP and TOC from sugarcane bagasse alkaline hydrolysate (lignin solution). This is an early report on statistical optimization of the removal of TP and TOC from sugarcane bagasse alkaline hydrolysate employing Fenton reagent mediated AOP process.

In chapter 12, we again turn to Brazil, where the authors discuss the decentralization of the Brazilian electricity sector in association with the internal electricity supply crisis, encouraging the sugarcane industry to produce electricity by burning sugarcane bagasse in cogeneration plants. This approach reduces the environmental impact of the sugarcane production and has opened up opportunities for distilleries and annex plants to increase their product portfolios. Potential scenarios for technically and environmentally improving the cogeneration performance were analyzed by using thermodynamic analysis and Life Cycle Assessment (LCA). The method used in this study aimed to provide an understanding and a model of the electrical and thermal energy production and the environmental impacts of conventional vapor power systems which operate with a Rankine cycle that are commonly used by Brazilian distilleries. Vapor power system experts have suggested focusing on the following technical improvement areas: increasing the properties of the steam from 67 bar and 480 °C to 100 bar and 520 °C, regeneration, and reheating. Eight case scenarios were projected based on different combinations of these conditions. A functional unit of "To the delivery of 1.0 MWh of electricity to the power grid from a cogeneration system" was defined. The product system covers the environmental burdens of the industrial stage and the agricultural production of sugarcane. Technical evaluation indicated that the energy efficiency improves as the pressure at which the vapor leaves the boiler increases. Simultaneously, the net power exported to the grid increases and the makeup water consumption in the cooling tower and the makeup water supplied to the boiler reduce. From the LCA, it was noted that the improved energy performance of the system is accompanied by reduced environmental impacts for all evaluated categories. In addition, vapor production at 100 bar and 520 °C results in greater environmental gains, both in absolute and relative terms. Reheating and regeneration concepts were found to be considerably effective in improving the energy and environmental performance of cogeneration systems by burning sugarcane bagasse. For the evaluated categories, the results indicate that the proposed modifications are favorable for increasing the efficiency of the thermodynamic cycle and for decreasing the environmental impacts of the product system.

In the final chapter, authors Petersen et al. compare the economics of producing only electricity from residues, comprised of surplus bagasse and 50% post-harvest residues, at an existing sugar mill in South Africa to the coproduction of ethanol from the hemicelluloses and electricity from the remaining solid fractions. Six different energy schemes were evaluated. They include:

1. exclusive electricity generation by combustion with high pressure steam cycles (CHPSC-EE)

2. biomass integrated gasification with combined cycles (BIGCC-EE)
3. coproduction of ethanol (using conventional distillation (CD)) and electricity (using BIGCC)
4. coproduction of ethanol (using CD) and electricity (using CHPSC)
5. coproduction of ethanol (using vacuum distillation (VD)) and electricity (using BIGCC)
6. coproduction of ethanol (using VD) and electricity (using CHPSC). The pricing strategies in the economic analysis considered an upper and lower premium for electricity, on the standard price of the South African Energy Provider Eskom' of 31 and 103% respectively and ethanol prices were projected from two sets of historical prices.

The authors found that from an energy balance perspective, ethanol coproduction with electricity was superior to electricity production alone. The VD/BIGCC combination had the highest process energy efficiency of 32.91% while the CHPSC-EE has the lowest energy efficiency of 15.44%. Regarding the economic comparison, the authors concluded that at the most conservative and optimistic pricing strategies, the ethanol production using VD/BIGCC had the highest internal rate of returns at 29.42 and 40.74% respectively. Ultimately, Petersen and his colleagues indicate that bioethanol coproduction from the hemicellulose fractions of sugarcane residues, with electricity cogeneration from cellulose and lignin, is more efficient and economically viable than the exclusive electricity generation technologies considered (at least under the constraints in found within a South African context).

PART I

WHY SUGARCANE?

CHAPTER 1

SUGARCANE BIOMASS PRODUCTION AND RENEWABLE ENERGY

MOSES ISABIRYE, D.V.N RAJU, M. KITUTU,
V. YEMELINE, J. DECKERS AND J. POESEN

1.1 INTRODUCTION

Bio-fuel production is rooting in Uganda amidst problems of malnutrition and looming food insecurity [1,2]. The use of food for energy is a worldwide concern as competition for resources between bio-fuel feedstocks and food crop production is inevitable. This is especially true for the category of primary feedstocks that double as food crops. Controversy surrounds the sustainability of bio-fuels as a source of energy in Uganda.

Given the above circumstances, adequate studies are required to determine the amount of feedstock or energy the agricultural sector can sustainably provide, the adequacy of land resources of Uganda to produce the quantity of biomass needed to meet demands for food, feed, and energy provision. Sugarcane is one of the major bio-fuel feed-stocks grown in Uganda.

Sugarcane Biomass Production and Renewable Energy. © *2013 Isabirye et al.* Biomass Now - Cultivation and Utilization, *edited by Miodrag Darko Matovic, ISBN 978-953-51-1106-1, Published: April 30, 2013 under CC BY 3.0 license.*

Growth in sugarcane cultivation in Uganda is driven by the increased demand for sugar and related by-products. Annual sugar consumption in Uganda is estimated at 9 kg per capita with a predicted per capita annual consumption increase by 1 % over the next 15 years [3].

This growth has resulted in increased demand for land to produce staple foods for households and thus encroaching on fragile ecosystems like wetlands, forests and shallow stoney hills and, a threat to food security [4]. The situation is likely to worsen with the advent of technology advancements in the conversion of biomass into various forms of energy like electricity and biofuels. A development that has attracted government and investors into the development of policies [5, 6] that will support the promotion of bio-fuels in Uganda [7].

Competition for land resources and conflicts in land use is imminent with the advent of developments in the use of agricultural crop resources as feedstocks for renewable energy production. Sugarcane is one such crop for which production is linked to various issues including the sustainability of households in relation to food availability, income and environmental integrity. The plans for government to diversify on alternative sources of energy with focus on biofuels and electricity generation has aggravated the situation. This chapter aims at demonstrating how sugarcane biomass can sustainably be produced to support fuel and electrical energy demands while conserving the environment and ensuring increased household income and food security.

This study was conducted with a major objective of assessing sustainable production of bio-fuels and electricity from sugarcane biomass in the frame of household poverty alleviation, food security and environmental integrity.

1.2 RESEARCH METHODS

The assessment of sugarcane production potential is done for the whole country. The rest of the studies were done at the Sugar estate and the out-grower farmers.

1.2.1 ASSESSMENT OF SUGARCANE PRODUCTION POTENTIAL

The overall suitability assessment involved the use of the partial suitability maps of temperature, rainfall and soil productivity ratings (Figures 1 and 2). An overlay of the three maps gave suitability ratings for sugarcane bio-fuel feedstock.

Subtraction of gazetted areas, wetlands and water bodies produced final suitability maps and tables presented in the results. Steep areas have not been excluded since they are associated with highlands which are densely populated areas. It is hoped that soil conservation practices will be practiced where such areas are considered for production of sugarcane feedstock. Urban areas, though expanding, are negligible and have not been considered in the calculations.

The suitability of the land resource quality for sugarcane was based on sets of values which indicate how well each cane requirement is satisfied by each land quality say: mean annual rainfall, minimum and maximum temperatures and soil productivity. The four suitability classes (rating), assessed in terms of reduced yields, and were defined according to the FAO [10]. Potential land-use conflict visualization also gives an indication of land available for the production of sugarcane bio-fuel feedstocks. Conflict visualization for food versus sugarcane was done by an overlay of suitability maps of maize with sugarcane. Land-use conflict with gazetted areas was assessed by overlaying gazetted area maps with sugarcane suitability map.

1.2.2 SUSTAINING SUGARCANE BIOMASS PRODUCTIVITY

The total biomass production of five commercial sugarcane varieties grown on the estate across all crop cycles (plant and three ratoons) was developed.

The data on cane yield and cane productivity of plant and ratoon crops between 1995-1996 and 2009-2010 were collected from annual reports and compared to experimental data. The total bio-mass (cane, trash and

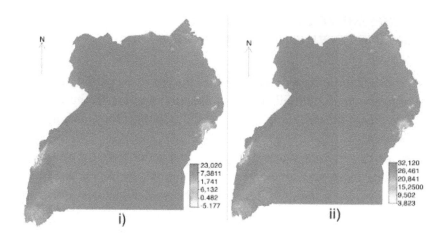

FIGURE 1: i) Minimum temperatures and maximum temperatures ii) "in [8]".

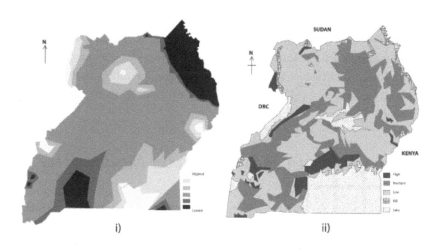

FIGURE 2: i) Mean annual rainfall " in [8] "; ii) Soil productivity ratings " in [9] ".

tops) production in plant and two ratoons were recorded at harvest age i.e. 18 and 17 months for plant and ratoon crops respectively. Data was collected from three locations of plot size 54 m² (4 rows of 10m length). Nutrient status of crop residues on oven dry basis was adopted as suggested by Antwerpen [11] for calculation of nutrient return to the soil and nutrients available to succeeding crop.

A replicated trial with four replications was established during 2009-2010 with different levels of chemical fertilizers and factory by-products (filter mud and boiler ash) to test their influence on cane and sugar yields.

Semi-commercial trials were established on the estate to study the influence of green manuring with sunn hemp against no green manured blocks (control), aggressive tillage against reduced tillage, and intercropping with legumes on cane yield and juice quality parameters.

Field studies were conducted during 2002-2004 and 2007-2009 to evaluate the influence of different levels of Nitrogen (N), Phosphorus (P), Potassium (K) and sulphur (S) on cane yield and juice quality of plant crop of sugarcane.

The cane yield data on green cane vs burnt cane harvesting systems, and aggressive vs reduced tillage operations were collected and analysed for biomass yield.

1.2.3 ASSESSMENT OF POTENTIAL BIOFUEL PRODUCTIVITY AND CANE BIOMASS ELECTRICITY GENERATION

Ethanol yield estimates from sugarcane is based on yield per ton of sugarcane. In addition, the production of bagasse from the cane stalk available for electricity generation were collected and analysed as per the following bagasse-steam-electrical power norms at Kakira sugar estate:

i. Bagasse production is 40 % of sugarcane production
ii. Moisture % in bagasse is 50%
iii. 1.0 ton of bagasse produces 2.0 tons of steam
iv. 5.0 tons of steam produces 1.0 Mwh electrical power

Therefore 2.5 tons of bagasse produces 5.0 tons of steam which will generate 1.0 Mwh electrical power. The electric power used in Kakira is hence generated from a renewable biomass energy source.

In 2005, Kakira had two 20 bar steam-driven turbo-generators (3 MW + 1.5 MW) in addition to 5 diesel standby generators. Thereafter, two new boilers of 50 tonnes per hour steam capacity at 45 bar-gauge pressure, with all necessary ancillaries such as an ash handling system, a feed water system and air pollution controls (such as wet scrubbers and a 40m 30 high chimney) were installed.

1.2.4 CONTRIBUTION TO HOUSEHOLD INCOME AND FOOD SECURITY

Indicative economic assessments included the use of gross sales for the raw material (farm gate) and ethanol. Annualized sugarcane net sales were compared to household annual expenditures to allow assessment of cane contribution to household income. Integration of commodity prices gives insight on the potential contribution of bio-fuels to household poverty alleviation and overall development of rural areas.

1.3 RESULTS

1.3.1 SUITABILITY

The agro-ecological settings favor the growing of sugarcane with a potential 10,212,757 ha (49.6%) at a marginal level of production with 2,558,698 ha (12.4%) land area potentially not suitable for cane production. Although the current production is far below the potential production [12], the related cane production is 908,935,330 and 60,769,069 tons respectively. It is also evident that there is possibility of increasing production through expansion of land area under sugarcane.

1. Suitability of sugarcane production and conflict visualization between food crops and gazetted areas in Uganda.

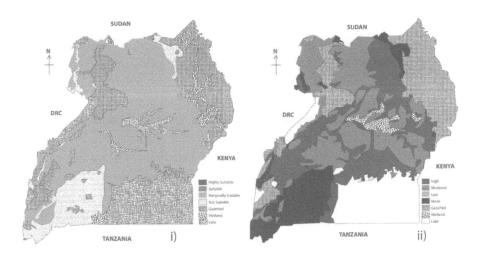

FIGURE 3: Sugar cane suitability ratings (i) and conflict visualization between food crops and gazetted areas.

The marginal productivity of cane in Uganda is a function of rainfall amount and the atmospheric temperature. Nevertheless the average optimum yields (89 ton / ha) at marginal level of productivity are comparable to yields of 85 ton / ha in a commercialized production in Brazil [13].

Expanding acrage under sugarcane is likely to increase pressure on gazetted biodiversity rich areas including wetlands with consequent potential loss of bio-diversity.

Sugarcane and maize (food crop) have similar ecological requirements, presenting a situation of high potential land-use conflict as 49.6 % of arable land can be grown with both sugarcane and food crops (figure 3 i). Figure 3 ii), shows 14 % of the land where sugarcane has potential conflict with gazetted areas of which 4.3 % has potential conflict with forest reserves.

Sugarcane, given its energy balance advantage, is likely to be beneficial if promoted as bio-fuel feedstock as this is likely to increase sugarcane prices to the benefit of the small scale farmer.

1.3.2 AGRONOMY

The beneficial effects of integrated agronomic practices like reduced till-age operations, balanced fertilization; organic recycling of mill by-prod-ucts (filter mud and boiler ash); intercropping with legumes; green manur-ing with sunn hemp; crop residue recycling through cane trash blanketing in ratoons by green cane harvesting to sustain soil fertility and cane pro-ductivity in monoculture sugarcane based cropping system are presented and discussed. Partitioning of dry matter between plant and ratoon crops of cane grown on the estate and outgrowers fields were quantified and also presented in this chapter.

1.3.2.1 INFLUENCE OF AGRONOMIC PRACTICES ON CANE YIELD AND CANE PRODUCTIVITY

1.3.2.1.1 GREEN MANURING

There was considerable increase in cane yield (7.92 tc/ha) and cane pro-ductivity (0.62 tc/ha/m) in plant and ratoon crops due to green manuring as compared to blocks without green manuring (Table 1).

Growing sunn hemp (*Crotolaria juncia*) during fallow period for in-situ cultivation has been a common practice to improve soil health on the estate since 2004. Sunn hemp at 50% flowering on average produces 27.4 t/ha and 5.9 t/ha of fresh and dry weights respectively. It contains 2.5% N on oven dry basis and adds about 147kg N/ha to the soil. Of this amount, 30% (44kg N/ha) is presumed to be available to the succeeding sugarcane plant crop. Antwerpen [11] reported that N available to sugarcane ranges between 30-60% of total N added to soils in South Africa.

1.3.2.1.2 BALANCED FERTILIZATION

The results indicated that application of N to plant crop at 100kg /ha, phos-phorus at 160kg P_2O_5 /ha, potassium at 100 K_2O /ha and sulphur at 40kg /ha significantly increased the cane yields by 23.3 tc/ha; 22.25 tc/ha; 12.07 tc/ha

TABLE 1: Cane yield and cane productivity variance due to green manuring.

Crop Cycle	Green Manuring		No Green Manuring		Variance	
	Yield tc/ha	Productivity tc/ha/m	Yield tc/ha	Productivity tc/ha/m	Yield tc/ha	Productivity tc/ha/m
Plant	125.3	5.9	112.2	5.2	13.1	0.8
Ratoon1	95.3	5.8	92.4	5.0	2.8	0.8
Ratoon 2	92.4	5.0	84.6	4.7	7.8	0.3
# average	104.3	5.6	96.4	5.0	7.9	0.6

and 8.71 tc/ha respectively over no application of N, P, K and S. Sugar yields were also improved due to N, P, K and S application by 2.91 ts/ha; 2.17 ts/ha; 2.88 ts/ha and 0.44 ts/ha respectively as compared to no application (Table 2).

TABLE 2: Influence of N, P, K and S nutrition on cane and sugar yields

Nutrient levels (kg/ha)	Attributes	
	Cane yield (tc/ha)	Sugar yield (ts/ha)
N levels: 0	112.68	15.69
50	124.38	17.08
100	135.98	18.60
P_2O_5 levels: 0	108.70	14.51
80	120.40	16.34
160	130.95	18.01
K_2O levels: 0	108.06	15.58
50	124.86	17.32
100	134.13	18.46
Sulphur levels: 0	113.90	11.92
40	122.61	12.36
CD at 5%: N	10.69	1.20
P_2O_5	9.80	1.18
K_2O	9.06	1.06
S	6.94	0.32

Balanced fertilizer application is very vital for crop growth. Adequate amounts of especially the major nutrients need to be supplied for proper crop growth. Excessive application of N in cane plant crop has been shown to inhibit the activity of free living N-fixing bacteria and chloride ions from Muriate of potash adversely affecting soil microbial populations [14].

1.3.2.1.3 MILL BY-PRODUCTS

Millable stalk population at harvest was significantly higher due to application of filter mud and boiler ash + 100% recommended dose of fertilizers (RDF) than all other treatment combinations. However, there was no significant effect on stalk length, number of internodes and stalk weight due to different treatments. Cane and sugar yields were significantly higher by 30.2 tc/ha and 3.8 ts/ha respectively due to application of filter mud and boiler ash + 100% RDF. The data are presented in Table 3.

Cane-mill by-products (filter mud and boiler ash) do contain valuable amounts of N, P, K, Ca, and several micronutrients [15]. These in addition to the inorganic fertilizers applied considerably increased cane yields as compared to treatments which received only inorganic fertilizers. In [11] it is also showed that organic wastes, including filter mud and bolier ash could be used as an alternative source of nutrients in cane cultivation.

TABLE 3: Influence of mill by-products on cane and sugar yields

Treatment	Yield (T/ha)	
	Cane	Sugar (Estimated))
1. 100% recommended dose of fertilizers (RDF)	131.6	16.6
2. Filter mud + Boiler ash @ 32 + 8 T/ha alone	122.1	16.2
3. 100% RDF+ Filter mud + Boiler ash @ 32 + 8 T/ha	161.7	20.4
4. 75% RDF+ Filter mud + Boiler ash @ 32 + 8 T/ha	143.8	17.5
5. 50% RDF+ Filter mud + Boiler ash @ 32 + 8 T/ha	142.1	17.3
6. 25% RDF+ Filter mud + Boiler ash @ 32 + 8 T/ha	135.0	16.6
CD @ 5%	12.3	1.9

1.3.2.1.4 PARTITIONING OF BIO-MASS

Among the crop cycles, plant crop recorded higher bio-mass production than succeeding ratoons. The production of crop residues were also higher (50.7 t/ha) in plant crop than 1st (45.7 t/ha and 2nd (36.6 t/ha) ratoons. The data are presented in the Table 4 below.

TABLE 4: Crop-wise partitioning of bio-mass

Crop cycle	Cane weight (Tc/ha)	Tops weight (T/ha)	Trash weight (T/ha)	Total bio-mass (T/ha)	Tops + Trash weight T/ha	% over total
Plant	137.5	30.3	20.4	188.2	50.7	27.0
Ratoon 1	124.9	26.8	18.9	170.59	45.7	27.0
Ratoon 2	107.5	19.7	16.8	144.02	36.6	25.4
# average	123.3	Fresh: 25.6	Fresh: 18.7	-	Fresh: 44.3	26.4
		Dry: 9.0	Dry: 16.9	-	Dry: 25.8	

Results indicate that on average, 25.8 tons of dry matter (cane trash and tops) is produced from each crop cycle at harvest. In burnt cane harvesting system, all this dry matter is lost unlike in green cane harvesting. This explains the gradual decline in cane yield in such harvesting systems. The decline is presumed to be due to deteriorating organic matter and other physical and chemical properties of the soil [11].

After decomposition of cane trash, 139kg N, 59kg P_2O_5, 745kg K_2O, 41kg Ca, 46kg Mg and 34kg S /ha were added to the soils and these added nutrients would be available to the succeeding crop at 30% of the total nutrients [11]. The nutrient concentration of crop residues (trash + tops) were taken into account for computing nutrient additions to the soil and their availability to the succeeding ratoon crops and the data are presented in Table 5.

TABLE 5: Nutrient status of crop residues and their availability to the succeeding crops

Nutrient status of crop residues (%)	Total dry matter (T/ha)	Total nutrients	
		Added to soil (kg/ha)	Available to crops @ 30% (kg/ha)
N : 0.54	25.8	139	42
P_2O_5 : 0.23	25.8	59	18
K_2O : 2.89	25.8	745	223
Ca : 0.16	25.8	41	12
Mg : 0.18	25.8	46	14
S : 0.13	25.8	34	10

1.3 RENEWABLE ENERGY POTENTIAL

1.3.1 ETHANOL PRODUCTIVITY

Sugarcane, given its energy balance advantage, is likely to be beneficial if promoted as bio-fuel feedstock as this is likely to increase sugarcane prices to the benefit of the small scale farmer.

Promoting sugarcane as a feedstock for ethanol is likely to improve rural livelihood and also minimize on forest encroachment since energy output per unit land area is very high for sugarcane.

In Brazil for example the production of sugar cane for ethanol only uses 1% of the available land and the recent increase in sugar cane production for bio-fuels is not large enough to explain the displacement of small farmers or soy production into deforested zones [13].

To minimize competition over land, it is advisable to grow sugarcane that has high yields with higher energy output compared to other biofuel crops. High yielding bio-fuels are preferable as they are less likely to compete over land [16].

1.3.2 BIOELECTRICITY GENERATION

Hydropower contributes about 90 per cent of electricity generated in Uganda with sugarcane based bagasse bioelectricity, fossil fuel and solar energy among other sources of power. Although the current generation of 800 MW [18]. has boosted industrial growth, the capacity is still lagging behind the demand that is driven by the robust growth of the economy.

The low pressure boilers of 45 bar currently generate 22 MW of which 10 MW is connected to the grid. However, Kakira sugar estate has a target of generating 50 MW of electricity with the installation of higher pressure boilers of 68 bar in 2013. This target can be surpassed given the abundance of the bagasse (Table 6 and 7).

TABLE 6: Mean estate cane production/productivity, electrical generation* norms (2008 – 2012).

Particulars	Plant	Ratoon 1	Ratoon 2	Ratoon 3	Total/ average
Cane harvest area (ha)	1,461.1	1,541.7	1,535.2	392.8	4,930.8
Total cane supply (tons)	155,207.3	162,132.3	137,436.4	40,138.9	494,915.9
Average cane yield (tc/ha)	106.23	105.16	89.52	87.28	100.37
Average harvest age (months)	19.20	18.15	17.98	16.50	17.96
Cane productivity (tc/ ha/m)	5.53	5.79	4.97	5.29	5.40
Bagasse production /ha	42.49	42.06	35.80	34.90	40.14
Steam generation (tons /ha)	84.98	84.12	71.60	69.80	80.28
Electric power generation (Mwh/ha)	17.00	16.82	14.32	13.96	16.05

* This electric power generation is calculated based on using low pressure boilers of 45 bar at Kakira estate

TABLE 7: Mean outgrowers cane production/productivity, electrical generation* norms (mean for 2008 – 2012).

Particulars	Plant	Ratoon 1	Ratoon 2	Ratoon 3	Total/ average
Cane harvest area (ha)	3,429.3	3,206.9	2,334.7	1,431.3	10,402.2
Total cane supply (tons)	320,100.20	290,073.29	188,274.65	113,837.25	912,285.49
Average cane yield (tc/ha)	93.34	90.45	80.64	79.53	87.70
Average harvest age (months)	18.50	17.50	18.00	16.00	17.50
Cane productivity (tc/ ha/m)	5.05	5.17	4.48	4.97	5.01
Bagasse production /ha	37.30	36.18	32.25	31.81	35.08
Steam generation (tons /ha)	74.60	72.36	64.50	63.62	70.16
Electric power generation (Mwh/ha)	14.90	14.50	12.90	12.70	14.00

*This electric power generation is calculated based on using low pressure boilers of 45 bar at Kakira estate
Tc = Tons of cane

Putting into consideration the productivity norms at Kakira estate and outgrowers (Table 8), with a potential of producing 908.9 m tons of sugarcane, Uganda has a potential of producing bio-electricity that surpasses the nation's demand by far. Much of this electrical power can be exported to the region, greatly expanding on Uganda's export base.

TABLE 8: Combined (Estate + Outgrowers cane production/productivity, Electrical power generation* norms (mean for 2008 – 2012)

Particulars	Plant	Ratoon 1	Ratoon 2	Ratoon 3	Total/ average
Cane harvest area (ha)	4,890.4	4,748.6	3,869.9	1,824.1	15,333.0
Total cane supply (tons)	475,307.50	452,205.60	325,711.05	153,976.15	1,407,200.4

TABLE 8: CONTINUED

Average cane yield (tc/ha)	97.19	95.22	84.17	84.41	91.78
Average harvest age (months)	18.75	17.75	18.00	16.00	17.65
Cane productivity (tc/ha/m)	5.18	5.36	4.67	5.27	5.20
Bagasse production /ha	38.88	38.09	33.67	33.76	36.71
Steam generation (tons /ha)	77.76	76.18	67.34	67.52	73.42
Electric power generation (Mwh/ha)	15.55	15.24	13.47	13.50	14.68

*This electric power generation is calculated based on using low pressure boilers of 45 bar at Kakira estate

1.3.2 HOUSEHOLD INCOME AND FOOD SECURITY

The competition for resources between sugarcane and food crops is apparent with foreseen consequent increased food insecurity. Fifty percent of the arable land area good for food crop production is equally good for sugarcane.

A farm household that allocates all of its one hectare of land to sugarcane is expected to earn 359 $ at high input level, 338 $ at intermediate level and 261 $ at low input level [4]. The 391 $ required to purchase maize meal is well above the net margins from one ha. This shows that proceeds from one hectare cannot sustain a household of 5. It is further revealed that maize produced from 0.63 ha can sustain a household nutritionally; however considering the annual household expenditure (760.8 $; [17]), about three hectares of land under sugarcane are required at low input level to support a household [4].

However, this study reveals that sugarcane sales accrued from ethanol under a scenario of a flourishing bio-fuel industry is associated with increased income that is likely to support households (Table 9). An ethanol gross sale per person per day is 1.6 dollars; an indication that the cultivation of sugarcane based biofuel is likely to contribute to alleviation of household poverty. A trickle-down effect on household income is expected from a foreseen expansion of bagasse-based electricity generation beyond the estate into the national electricity grid.

TABLE 9: Sugarcane productivity, sales and potential land-use conflict

Production / year		Gross sales			Conflict		
Cane production	Billion	Farm	Ethanol	Capita	Food	Gazetted	Forest
ton	litres	/ha/year			%		
908.9 m	75.4	1869	22161	1.6	50.0	14.0	4.3

Sugarcane= USD 21/ton: projected population of 33 m in 2009 is used

1.4 FURTHER RESEARCH

The expansion of cane production is largely driven by market forces oblivious to the detrimental impact the industry is likely to have on food, livelihood security and the status of biodiversity. In addition to lack of appropriate policies to support the small-scale cane farmer, the policies are largely sectoral with no linkages with other relevant policies. Information is required to support the sustainable development of the cane industry with minimal negative impact on food and livelihood security and the status of biodiversity.

1.5 RELEVANT QUESTIONS TO EXPLORE

Can food crop productivity be improved in the context of a sugarcane-based farming system?

Can the understanding of the dimensions of food and livelihood security in sugarcane-based farming systems inform the synergistic development and review of relevant policies in the food, agriculture, health, energy, trade and environment sectors? What are the social impacts of the industry in light of the various agro-ecological zones of the country? What is the gender based livelihood strategies with special emphasis on labor exploitations, child labor etc?

What do people consider as possible options for improving food and livelihood security in a sugarcane-based farming system? Do these options differ between different actors (local women and men, NGOs and government)? How do families cope with food inadequacy, inaccessibility and malnutrition?

Can the study inform the carbon credit market initiative for farming systems in Uganda through the climate smart agriculture concept? Are the proposed assessment tools appropriate for Ugandan situations and the cane-based systems in particular?

1.6 CONCLUSION

Driven by the need to meet the increasing local and regional sugar demand, and fossil fuel import substitution, cane expansion has potential negative impact on food security and biodiversity. However, this negative impact parallels the benefits related to cane cultivation. Cane biomass yield can be improved and sustained through the integrated use of various practices reported in this study. Consequently this reduces the need to expand land acreage under cane while releasing land for use in food crop productivity. The high biomass returned to the ground sequesters carbon thereby offering the opportunity for sugarcane based farmers to earn extra income through the sale of carbon credits. Trickle down effects are expected to increase household income through the production and marketing of cane based biofuel and electricity.

These developments are expected to improve the farmers purchasing power, making households to be less dependent on the land and more food secure financially.

REFERENCES

1.   B. A Bahiigwa, Godfrey, (1999) Household Food Security In Uganda: An Empirical Analysis. Economic Policy Research Center, Kampala, Uganda

2. Uganda government (2002) Uganda Food and Nutrition Policy, Ministry of Agriculture, Animal Industry and Fisheries, Ministry of Health, Kampala, Uganda

3. USCTA (2001) The Uganda Sugarcane Technologist's Association, Fourth Annual Report, 2001, Kakira, Uganda.

4. Isabirye (2005) Land Evaluation around Lake Victoria: Environmental Implications for Land use Change. PhD Dissertation, Katholieke Universiteit, Leuven, Belgium

5. MEMD (2007) The Renewable Energy Policy for Uganda. MEMD, Kampala, Uganda.

6. MEMD (2010) The Uganda Energy Balance Report. MEMD, Kampala, Uganda.

7. Bio-fuel-news (2009) Uganda to produce cellulosic ethanol in a year. Bio-fuel International 310 http://www.Bio-fuel-news.com/magazine_store.php?issue_id=34.

8. Meteorology Department (1961) Climate data. Meteorological Department, Entebbe Uganda.

9. Chenery (1960) Introduction to the soils of the Uganda Protectorate, Memoirs of the Research Division, Series 1, Soils, 1, Department of Agriculture, Kawanda Research Station, Uganda.

10. FAO (1983) Guidelines: Land evaluation for rainfed agriculture. FAO Soils Bulletin 52, Food and Agricultural Organization of the United Nations, Rome.

11. R. V Antwerpen, (2008) Organic wastes as an alternative source of nutrients. The link published by SASRI. 172 May 2008. 8-9.

12. FAOStat (2012) http://faostatfao.org/site/339/default.aspx Sunday, May 06, 2012.

13. M. R Xavier, (2007) The Brazilian sugarcane ethanol experience. Issue Analysis, 3, Washington, USA, Competitive Enterprise Institute. 11 p.

14. A.P, Carr, D.R, Carr, I.E, Wood, A.W and Poggio, M. (2008) Implementing sustainable farming practices in the Herbert: The Oakleigh farming company experience

15. D. V. N Raju, and K. G. K Raju, (2005) Sustainable sugarcane production through integrated nutrient management. In: Uganda Sugarcane Technologists' Association 17th Annual Technical Conference.

16. Pesket Leo, Rachel Slater, Chris Steven, and Annie Dufey (2007) Biofuels, Agriculture and Poverty Reduction. Natural Resource Perspectives 107, Overseas Development Institute, 111 Westminster Bridge Road, London SE1 7JD.

17. UBOS (2001) Uganda National household survey 1999/2000; Report on the socio-economic. Uganda Bureau of Statistics, Entebbe, Uganda. www.ubos.org.

18. Ibrahim Kasita (2012) Strategic plan to increase power supply pays dividends. New Vision, Publish Date: Oct 09, 2012.

CHAPTER 2

EVALUATING THE COMPOSITION AND PROCESSING POTENTIAL OF NOVEL SOURCES OF BRAZILIAN BIOMASS FOR SUSTAINABLE BIORENEWABLES PRODUCTION

MARISA. A LIMA, LEONARDO D. GOMEZ,
CLARE G. STEELE-KING, RACHAEL SIMISTER,
OIGRES D. BERNARDINELLI, MARCELO A. CARVALHO,
CAMILA A. REZENDE, CARLOS A. LABATE,
EDUARDO R. DEAZEVEDO, SIMON J. MCQUEEN-MASON AND
IGOR POLIKARPOV

2.1 BACKGROUND

The production of biorenewables, particularly liquid biofuels, from ligno-cellulosic biomass has become a strategic research area because it holds the potential to improve energy security, decrease urban air pollution and reduce CO_2 accumulation in the atmosphere [1,2]. In turn, the biorefining platforms required for biofuels production present an opportunity to stimulate new markets for the agriculture sector and increase domestic employment, contributing to the development of emerging economies [3].

Evaluating the Composition and Processing Potential of Novel Sources of Brazilian Biomass for Sustainable Biorenewables Production. © *2014 Lima et al.; licensee BioMed Central Ltd.* Biotechnology for Biofuels *2014, 7:10 doi:10.1186/1754-6834-7-10. Creative Commons Attribution License (http:// creativecommons.org/licenses/by/2.0).*

In Brazil, the production of first-generation ethanol from sugarcane juice (sucrose) has made the country a leading producer of biofuels. At present, approximately 90% of the automobiles made in Brazil are dual-fuel [4]. In 2006 to 2007, Brazilian ethanol production reached 18 billion liters, supplying the domestic demand and producing an excess of 3.5 billion liters for export. International targets for a reduction in CO_2 emission combined with high oil prices are driving an increase in global bioethanol production, which is predicted to reach 43 billion liters in 2025. Meeting this demand will require a 130% increase in the area of cultivated sugarcane [5]. In this context, second-generation biofuels, which use, for example, biomass feedstocks, agricultural wastes and wood residue, represent an efficient and complementary approach to increase liquid biofuel production. The adoption of second-generation bioethanol production from lignocellulosic biomass is attractive from a number of perspectives. By making use of all available biomass, such approaches can improve the carbon footprint of biofuels further, as well as increasing the yield of ethanol per hectare and providing a means to sustain the operation bioethanol plants throughout the year, instead of their current seasonal operation [2,6,7].

The diversity of climates and agricultural conditions in Brazil enables the growth of a large diversity of lignocellulosic materials. The management of this primary productivity can be driven towards high output/low input systems, which are optimal for second-generation fuels. In addition, Brazilian agriculture provides large volumes of lignocellulosic residues that could be used for biofuel production.

Among these residues, sugarcane bagasse is the most promising Brazilian feedstock for lignocellulosic ethanol production, being a by-product of first-generation ethanol production and therefore available in large amounts at sugarcane mills. According to the Brazilian Ministry of Agriculture, sugarcane production for 2012 to 2013 is estimated to reach 650 million tons and each ton of cane milled generates approximately 260 kg of bagasse, which is used mainly to co-generate the electricity needed for the operation of the mill [8,9]. Thus, to date, most research has focused on sugarcane bagasse as a feedstock for second-generation biofuel production, with the potential to increase bioethanol production in Brazil by one third.

However, increasing Brazilian bioethanol production by one third will be insufficient to meet future demand, and it is clear that consideration of other sources of biomass is necessary. Brazil has around 6.5 million hectares of cultivated forest, among which 4.8 million hectares are occupied by eucalyptus and the remaining fraction by pine. The forest industry is a source of large quantities of lignocellulosic residues such as bark and branches, which can potentially be used for second-generation bioethanol, but are currently left in the field [10,11]. Approximately 30% of the total biomass produced in Brazil by eucalyptus forestry is lost as residues, when the trees are harvested at the end of a seven-year cycle. The bark proportion in eucalyptus forestry can reach between 10% and 12% of the total biomass harvested, which represents a volume of 15 to 25 ton/ha/year [12-14], making this a promising feedstock for bioethanol production [15].

The diversification of feedstock for lignocelluloses-derived fuels requires an innovative approach that expands beyond the agricultural wastes. Perennial grasses, such as miscanthus and switchgrass, have been proposed as key bioenergy crops in Europe and the US, based on their low input and marginal land requirements. These biomass grasses could also make a substantial contribution within the Brazilian energy matrix, serving as an alternative to sugarcane inter-season, when there is no bagasse production. Although switchgrass and miscanthus could be used in Brazil, there are also a number of other candidate biomass grasses that are already established and characterized from an agronomical point of view. Brazil has the fourth largest worldwide cultivated pasture area, reaching around 174 million hectares; around 30% of national territory, distributed throughout the country [16,17]. The tropical climate in Brazil supports the efficient growth of a range of grasses with high productivity, for example, from the genus *Brachiaria*, *Panicum*, *Pennisetum* and *Cynodon*, which are very important for Brazilian beef and dairy cattle production. Brachiaria was first introduced to Brazil 15 years ago, and today occupies around 70% of total pasture area, followed by *Panicum*, which occupies approximately 10%. Initial studies have shown promising averages of productivity yields (dry mass) for different perennial grasses species compared to sugarcane, for example: *Pennisetum purpureum* (35 ton/ha), *Panicum maximum* (30 ton/ha) and *Brachiaria brizantha* (20 ton/ha), compared to sugarcane at 30 ton/ha [18].

In this paper, we have investigated and compared the potential of three grasses (*Panicum maximum, Pennisetum purpureum* and *B. brizantha*) and eucalyptus barks (from *Eucalyptus grandis* and the hybrid *E. grandis x urophylla*) against sugarcane bagasse as feedstocks for bioethanol production. We examined the general composition of these potential feedstocks and compared their suitability for processing to produce sugars for fermentation under a range of conditions. The aim of this characterization was to increase the range of potential feedstocks for Brazilian biofuel production to include sustainable biomass sources outside the human food chain.

2.2 RESULTS AND DISCUSSION

The development of second-generation biofuels requires a diverse set of feedstocks that can be grown sustainably and processed cost effectively. In particular, many biofuel production plants operate seasonally and stand idle for several months of the year, and this is unsatisfactory as it denotes an inefficient use of capital as well as providing only intermittent employment for workers. One way to avoid discontinuous biofuel production is to use a wider range of biomass sources that may be available during the current idle periods. Here, the potential of three widely grown, high-yielding Brazilian grasses, as well as the bark from two commercial eucalyptus clones, was investigated and compared with sugarcane bagasse, the most widely used biomass for bioethanol production. The biomasses were subjected to a range of pretreatment conditions to evaluate their effects on cellulose accessibility and enzymatic digestibility, as well as the levels of inhibitors produced.

2.2.1 BIOMASS COMPOSITION

For all six feedstocks, the biomass composition was analyzed for soluble extractives, silicon, cellulose, hemicellulose and lignin contents (Table 1). Bagasse is extensively washed during the commercial extraction of sucrose for first-generation ethanol and, as expected, the sequential extraction using organic solvents revealed a lower soluble content in sugar-

TABLE 1. Biomass composition of raw Brazilian biomasses.

Biomass	Solubles (%)	Silicon (%)	Cellulose (%)	Hemicellulose (%)	Lignin (%)	Maximum theoretical ethanol yield (L/dry ton)[a]	Productivity (ton/ha)[b]	Maximum theoretical ethanol yield (L/ha)[c]
Sugarcane bagasse	3.39 ± 1.26	0.44 ± 0.03	39.44 ± 1.21	27.45 ± 2.08	27.79 ± 1.39	282.62	30	8,478.6
Panicum maximum	5.23 ± 2.37	1.07 ± 0.01	39.87 ± 1.97	26.62 ± 1.46	25.36 ± 1.06	285.70	30	8,571.0
Pennisetum purpureum	5.70 ± 2.25	0.85 ± 0.01	45.97 ± 3.10	27.03 ± 1.02	22.80 ± 1.26	329.41	35	11,529.4
Brachiaria brizantha	12.41 ± 3.69	1.38 ± 0.06	43.48 ± 1.84	23.23 ± 3.16	23.09 ± 0.73	311.57	20	6,231.4
E. grandis bark	28.29 ± 3.43	0.03 ± 0.01	39.54 ± 1.10	18.84 ± 4.11	21.57 ± 1.59	283.34	25	7,083.5
E. grandis x urophylla bark	28.13 ± 2.20	0.03 ± 0.01	40.36 ± 4.31	16.45 ± 3.05	22.18 ± 2.22	289.21	25	7,230.3

[a] Calculated considering the total cellulose conversion in the sample, according to the National Renewable Energy Laboratory standards [19].

[b] Source: [11,18].

[c] Calculated with base on total cellulose conversion in the sample and average Brazilian biomasses productivity (ton/ha).

cane bagasse (3.39 ± 1.26%). *Panicum maximum* and *Pennisetum purpureum* showed 5.23 ± 2.37% and 5.70 ± 2.25% of solubles, respectively, whereas *B. brizantha* had more than twice as much soluble material (12.41 ± 3.69%) as all three of these feedstocks. The amount of solubles extracted from eucalyptus bark (approximately 27%) was much higher, which correlates with previous results published by our research group [15].

Silicon is considered an important macronutrient for plant growth and development, particularly in grasses, where it is important for tissue strength and resistance to environmental stress and pathogens [20]. Generally, silicon represents the major mineral content in grasses and can accumulate up to 15% in some species such as rice, where it mostly occurs as amorphous silica with some silicon dioxide [21]. Silicon can cause problems in certain industrial processes [22,23], so it is pertinent to assess silicon levels in potential biomass sources. Quantification of silicon by X-ray fluorescence (XRF) shows that the perennial grasses, *B. brizantha* (1.38 ± 0.06%), *Panicum maximum* (1.07 ± 0.01%) and *Pennisetum purpureum* (0.85 ± 0.01%) contain higher silicon levels than sugarcane bagasse (0.44 ± 0.03%) (Table 1), whereas silicon levels in bark were much lower (0.03 ± 0.01 for both clones). The inorganic fraction of eucalyptus barks is composed mainly of calcium crystals in the form of calcium oxalate or carbonate [24,25]. The higher amount of silicon in the perennial grasses was accompanied by the presence of phytoliths, classified as panacoids, on the biomass surface, as observed by scanning electron microscopy (Additional file 1). Phytoliths are microscopic silica bodies that precipitate in or between cells of living plant tissues and are especially abundant, diverse and distinctive in the grass family [26].

Levels of cellulose, hemicellulose and lignin were determined biochemically and the results are shown in Table 1. Lignin is a complex polymer of phenyl propane units (p-coumaryl, coniferyl and sinapyl alcohol) that acts as a cementing and waterproofing agent. It is generally considered to be a barrier to the efficient saccharification of biomass [27].

Lignin content varied from 27.79% in sugarcane bagasse to approximately 22% in eucalyptus bark, with intermediate values in the perennial grasses. The hemicellulose fraction of the feedstocks was higher in the grasses, varying from 27% in sugarcane bagasse to 23% in *B. brizantha*, and was considerably lower in eucalyptus bark at about 19% and 16% for

SUPPLEMENTARY FILE 1

Surface images obtained by SEM showing the silica bodies (phytoliths) on plant tissue. (a) *P. maximum*; (b) *P. purpureum* and (c) *B. brizantha*.

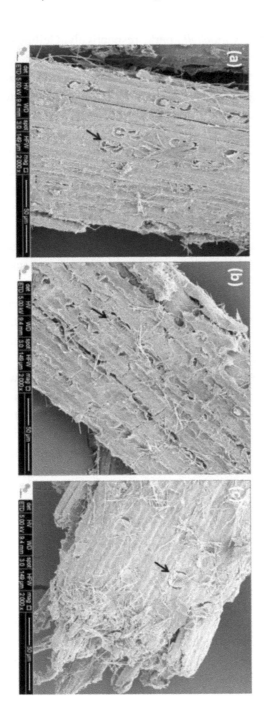

E. grandis and *E. grandis x urophylla* bark, respectively. Cellulose content, on the other hand, was highest in *Pennisetum purpureum* (46%), followed by *B. brizantha* (43%), whereas sugarcane bagasse, *Panicum maximum* and both eucalyptus barks showed a cellulose content of approximately 40%.

The carbohydrate fraction of these biomasses represents their potential for the biochemical conversion of sugars into lignocellulosic ethanol. Using the standard equations from the National Renewable Energy Laboratory [19] and considering total conversion of the cellulosic fraction, the potential ethanol yield (L/dry ton) for each biomass was calculated and is presented on Table 1. The highest ethanol yield (329.41 L/dry ton) was found for *P. purpureum*, reflecting its high cellulose content. Considering the biomass productivity values taken from published literature (also shown on Table 1) [11,18], it was possible to estimate the total theoretical ethanol yield (L/ha) for each of the evaluated feedstocks. It must be emphasized that these are simple approximate indications because the data are derived from a range of different crop yield studies and based on theoretical fermentation yield calculations. However, these values suggest that *Pennisetum purpureum* looks particularly promising due to its higher biomass productivity and cellulose content (around 35 ton/ha), which suggests a theoretical ethanol yield of more than 11,500 L/ha. This compares favorably with the first generation Brazilian bioethanol productivity from sugarcane juice, at around 6,000 L/ha [28]. As has been previously discussed, the yield of ethanol from bark could be higher than reported here, as considerable amounts of sugar occur in the soluble extractives (not included in this calculation), but this depends on how soon after harvest the bark is processed [29].

2.2.2 IMMUNOLABELING OF HEMICELLULOSE POLYSACCHARIDES

The composition of the hemicellulosic fraction of a biomass feedstock is one of the key determinants in selecting a choice of process for conversion. Paradoxically, the C5 sugars present in hemicelluloses represent both a hurdle for fermentation and a source of platform chemical for added value products. A rapid and reliable way to evaluate the relative content of key polysaccharides in the hemicellulosic fraction is by using immunobased techniques. Here, we used an ELISA-based approach to compare the six biomasses for their xylan, arabinoxylan, mannan, galactomannan, and glucomannan content. The hemicellulosic fraction was extracted with sodium hydroxide and analyzed by ELISA using the following antibodies: LM10 (recognizes unsubstituted and relatively low-substituted xylans, and has no cross-reactivity with wheat arabinoxylan), LM11 (recognizes unsubstituted and relatively low-substituted xylans, but can also accommodate more extensive substitution of a xylan backbone and binds strongly to wheat arabinoxylan) and LM21 (binds effectively to β-$(1 \rightarrow 4)$-manno-oligosaccharides from DP2 to DP5, displays a wide recognition of mannan, glucomannan and galactomannan, and has no known cross-reactivity with other polymers) [30-32]. Figure 1 shows that the hemicellulose fraction from the grasses gave strong signals with LM10 and 11 antibodies indicating a high content of xylans and arabinoxylans as typically seen in grasses, with lower signals for the mannan-detecting LM21 antibody. The hemicellulose fraction of sugarcane bagasse, *Panicum maximum*, *Pennisetum purpureum* and *B. brizantha*, after an initial 40-times dilution, showed a relative absorbance more than 12 times higher than the absorbance found for the positive control (10 µg/mL xylan). By contrast, xylan levels appeared lower in the hemicelluloses fraction from eucalyptus barks at the same initial dilution. The relative absorbance for the barks was around 2.5 times the positive control when LM10 was used.

The hemicellulose fractions from all biomasses were also diluted 40 times before the immunolabeling assays using LM11. However, the relative absorbance found for the grasses were reduced by approximately half (around 5.7 times the positive control), indicating lower arabinoxylan content when compared to xylans. By contrast, the relative absorbance for the eucalyptus barks increased by around 3.2 times, suggesting a higher content of derived arabinoxylans on its hemicellulose fraction.

When LM21 was used, the initial hemicelluloses fractions dilution needed was only 20 times, indicating a lower content of mannans polysaccharides for all six feedstocks when compared to xylans and arabinoxylans. The relative absorbance found for the three grasses were lower (around 0.4 times) than the positive control (galactomannan, 10 µg/mL),

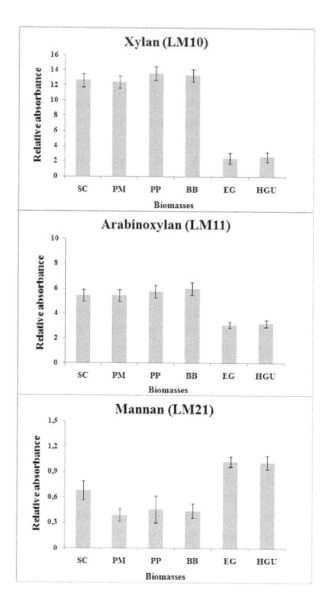

FIGURE 1: ELISA of xylans (LM10 and LM11) and mannan (LM21) polysaccharides on hemicellulose fraction from Brazilian grasses and Eucalyptus barks. SC, sugarcane bagasse; BB, *Brachiaria brizantha*; EG, *Eucalyptus grandis* bark; HGU, bark of hybrid between *Eucalyptus grandis x urophylla*; PM, *Panicum maximum*; PP, *Pennisetum purpureum*. Values expressed as relative absorbance to the positive control (xylan - μg/mL).

while for both Eucalyptus barks it was approximately the same as the control. The relative absorbance for sugarcane bagasse was 0.7 times that of the positive control. The higher relative absorbance found for eucalyptus barks suggests a higher content of mannans compared to the grasses and sugarcane bagasse.

The hemicellulose fractions from all biomasses were also diluted 40 times before the immunolabeling assays using LM11. However, the relative absorbance found for the grasses were reduced by approximately half (around 5.7 times the positive control), indicating lower arabinoxylan content when compared to xylans. By contrast, the relative absorbance for the eucalyptus barks increased by around 3.2 times, suggesting a higher content of derived arabinoxylans on its hemicellulose fraction.

When LM21 was used, the initial hemicelluloses fractions dilution needed was only 20 times, indicating a lower content of mannans polysaccharides for all six feedstocks when compared to xylans and arabinoxylans. The relative absorbance found for the three grasses were lower (around 0.4 times) than the positive control (galactomannan, 10 µg/mL), while for both Eucalyptus barks it was approximately the same as the control. The relative absorbance for sugarcane bagasse was 0.7 times that of the positive control. The higher relative absorbance found for eucalyptus barks suggests a higher content of mannans compared to the grasses and sugarcane bagasse.

2.2.3 EFFECT OF PRETREATMENTS ON THE COMPOSITION OF DIFFERENT FEEDSTOCKS

There is consensus regarding the need for a pretreatment to remove and/ or modify the matrix of lignin and hemicellulose surrounding the cellulose fraction, to enable efficient enzymatic saccharification of cellulose [33]. However, the complexity and heterogeneity found in the lignocellulosic biomass of different species makes it is advisable to optimize a pretreatment for each feedstock, to enable maximum saccharification whilst avoiding the generation of inhibitors of fermentation, such as furfurals. Ideally, a pretreatment should preserve the hemicellulose fraction, limit inhibitor formation, minimize the energy input, be cost-effective, warrant the recovery of high value-added co-products (for example, lignin) and minimize the production of toxic waste [12,34].

Since the composition of different biomasses affects the efficiency of processing, it will also influence the choice of pretreatments required to maximize the recovery of sugars. To evaluate this particular issue, we pretreated the six feedstocks under acid, alkaline, sulfite and hot water conditions over a range of temperatures. Figure 2 shows the averages of the three main components (cellulose, hemicellulose and lignin) content determined using different methods at microscale, as described in the Materials and Methods section. The standard deviations found for each of three components of the biomasses are also given in Figure 2.

Hot water pretreatment showed a similar effect over the chemical composition of the different biomasses, removing mainly the hemicellulose fraction. The lignin content remained fairly constant (varying between 27% and 23%), while the average cellulose content increased from around 40% to 60% as the temperature increased to 130°C (Figure 2). This enrichment in cellulose is a direct consequence of the removal of hemicellulose. However, at 180°C, the cellulose content was lower, possibly due to the production of degrading compounds such as furaldehydes, rather than a reduction in hemicelluloses removal at this temperature. On average, pretreatment at 180°C resulted in a reduction in the hemicelluloses fraction from approximately 25% (untreated feedstocks) to 13% (pretreatment at 180°C), ranging between $10.74 \pm 0.62\%$ in *B. brizantha* and $15.09 \pm 1.08\%$ in *E. grandis* bark (Figure 2).

The acid pretreatment was highly efficient for hemicellulose removal, and an increase in temperature (up to 130°C) had a further positive effect when compared to hot water treatment. However, at 180°C, the degrading hemicellulose product, 2-furfuraldehyde, was detected for all three grasses, with a higher content in *B. brizantha* liquor fraction, and for *E. grandis* bark. At the highest temperature (180°C), higher cellulose losses were also observed, and the average cellulose content decreased to around 60% after acid pretreatment at 180°C, compared to 70% at 130°C. However, even with the increase in temperature, acid pretreatment was not sufficient for lignin removal (Figure 2). At the highest temperature applied in this study (180°C), approximately 20% of remaining dry matter was lignin.

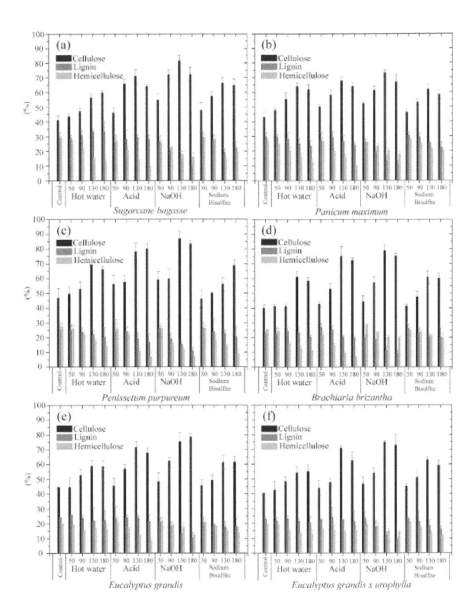

FIGURE 2: Chemical composition of non-pretreated and pretreated biomasses. (a) Sugarcane bagasse; (b)*Panicum maximum*; (c)*Pennisetum purpureum*; (d)*Brachiaria brizantha*; (e)*Eucalyptus grandis* bark; (f) bark of *E. grandis x urophylla*. Pretreatment types and temperatures are indicated.

The highest cellulose enrichment was observed in samples subjected to the alkaline pretreatment using sodium hydroxide, which removed higher quantities of both lignin and hemicellulose fractions. The average lignin content across all feedstocks was reduced from around 27% to 9% at 180°C. However, at this temperature, some cellulose losses were observed, particularly in sugarcane bagasse and *Panicum maximum*.

The chemical composition of biomasses submitted to treatment with sodium bisulfite at increasing temperatures was observed to be similar to hot water pretreatment. In all feedstocks, an increase in cellulose enrichment was observed until 130°C, reaching around 60%, with a maximum enrichment observed for sugarcane bagasse (66.1 ± 4.14%). At 180°C, a slight decrease in cellulose content was observed for sugarcane bagasse, *Panicum maximum* and the bark of *E. grandis x urophylla*. The cellulose fraction from *B. brizantha* and *E. grandis* remaining constant between 130°C and 180°C. Conversely, *Pennisetum purpureum* showed a gradual increase on its cellulose fraction until 180°C, reaching around 56% at 130°C and 68% at the highest temperature. The discrete cellulose enrichment observed after sulfite pretreatment is associated with a low removal of both hemicellulose and lignin.

The content of amorphous and crystalline cellulose after different pretreatment conditions was determined by a chemical method and each fraction is shown in Figure 3. We observed a clear increase in the crystalline portion of the cellulosic fraction until 130°C for all species and all pretreatments used. At 180°C, however, some losses in the crystalline fraction could be observed, mainly after hot water and acid pretreatment for the grasses. Analysis of the amorphous content of control samples indicated a variation of between 2% and 13% of total cellulose content in this fraction. The highest amorphous content was observed for *Pennisetum purpureum* (approximately 6% of cell-wall composition), followed by sugarcane bagasse (about 5%). The lowest amorphous cellulose content was observed in *B. brizantha*. No clear correlation between pretreatment conditions and the amorphous cellulose fraction was determined. However, considering the glucose content in the soluble fraction from pretreatment, it is possible that at lower temperatures this fraction was mainly removed, while at higher temperatures there was also a degree of biomass amorphization.

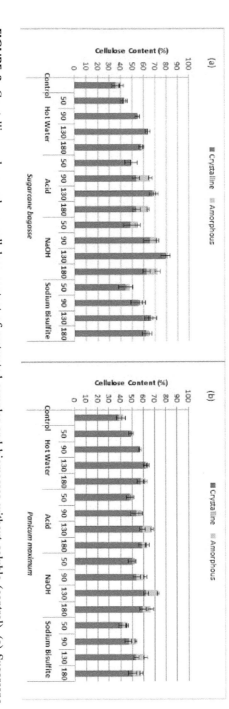

FIGURE 3: Crystalline and amorphous cellulose content of pretreated samples and biomasses without soluble (control). (a) Sugarcane bagasse; (b)*Panicum maximum*; (c)*Pennisetum purpureum*; (d)*Brachiaria brizantha*; (e)*Eucalyptus grandis* bark; (f) bark of *E. grandis* x *urophylla*. Pretreatment types and temperatures are indicated.

Sugarcane as Biofuel Feedstock

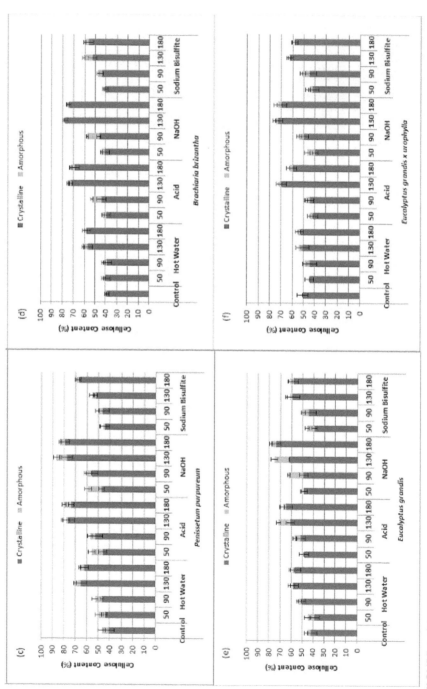

FIGURE 3: CONTINUED.

Hemicellulose fractions were analyzed after pretreatment to evaluate the changes in monosaccharide composition (Figure 4). Sugarcane bagasse, *Panicum maximum*, *Pennisetum purpureum* and *B. brizantha* showed a similar composition in the hemicellulose fraction, composed mainly of xylose, arabinose and glucose, followed by lower amounts of galactose and fucose. The hemicellulose fraction from eucalyptus barks was more heterogeneous, with lower xylose content when compared to the grasses (Figure 4). Barks showed a high amount of mannose and rhamnose, not detected in the grasses. These results were in agreement with the ELISA results, which indicated a higher content of xylans and arabinoxylans in the grasses and a significant level of mannans in the eucalyptus bark feedstocks.

2.2.4 SOLID-STATE NUCLEAR MAGNETIC RESONANCE

The effect of pretreatment on the feedstock compositions was also investigated using solid-state nuclear magnetic resonance (NMR). Figure 5 shows cross-polarization under magic angle spinning with total suppression of spinning sidebands (CPMASTOSS) spectra of the solid fractions of sugarcane bagasse samples submitted to the different pretreatments, which was very similar in all three novel grasses. All spectra were normalized with respect to line 10 (C1 carbon of cellulose). Chemical shift assignments based on the comparison with previously reported 13C NMR spectra of wood [35,36] and sugarcane bagasse [37] are listed in the caption of Figure 5 (see more complete attributions in table two of reference [37]).

The spectra of samples pretreated with hot water (Figure 5a) or sodium bisulfite (Figure 5b) at different temperatures were all similar to that of the untreated sample, showing that these pretreatments did not promote the efficient removal of hemicellulose and lignin, which is in agreement with the chemical composition analysis. Samples pretreated with sulfuric acid at temperatures up to 90°C also presented similar spectra to the untreated sample (Figure 5c). However, for pretreatment temperatures above 130°C there was a clear reduction in the hemicelluloses signals, lines 1 and 17, with little alteration in the lignin signals, lines 2, 11, 12, 13, 14 and 15.

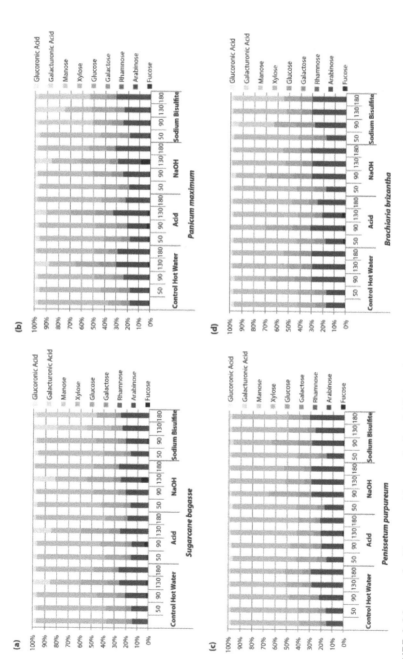

FIGURE 4: Monosaccharide composition on the hemicellulose fraction of pretreated samples and biomasses without soluble (control). (a) Sugarcane bagasse; (b)*Panicum maximum*; (c)*Pennisetum purpureum*; (d)*Brachiaria brizantha*; (e)*Eucalyptus grandis* bark; (f)*E. grandis x urophylla* bark. Pretreatment types and temperatures are indicated.

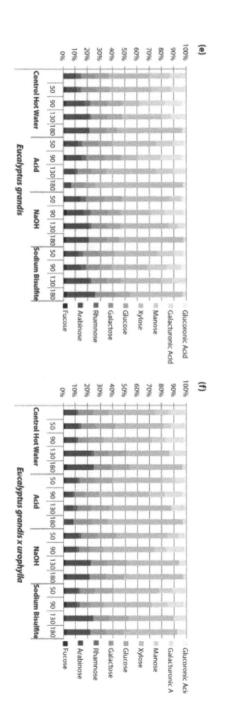

FIGURE 4: CONTINUED.

This suggests that, for sugarcane bagasse and the grasses, the pretreatment using sulfuric at 130°C acid is already effective for hemicellulose removal, but does not have a significant effect on lignin content. In samples pretreated with sodium hydroxide (Figure 5d), hemicellulose signals were already absent at 50°C whereas the lignin signals were reduced in line with an increase in pretreatment temperature. Indeed, the relative lignin content in the samples appeared similar for pretreatments at 130°C and 180°C, which suggests that the sodium hydroxide pretreatment of sugarcane bagasse and grasses at 130°C might be sufficient for the removal of hemicelluloses as well as effecting a reduction in lignin content.

NMR can also be used to give an indication of the composition of the crystalline cellulose to amorphous fraction after alkaline pretreatment. In feedstocks pretreated with sodium hydroxide at the higher pretreatment temperature (180°C), a decrease in the intensity ratio between lines 3 and 4 as well as between lines 7 and 8 was observed, which may be interpreted as a consequence of the removal of amorphous cellulose content by pretreatment.

NMR measurements were also carried out for the other grass feedstocks after distinct pretreatments and exhibited a response similar to that of sugarcane bagasse (data not shown). Conversely, NMR studies of the two types of eucalyptus bark show some particularities. Figure 6 shows CPMASTOSS spectra of the solid fractions of *E. grandis x urophylla* samples submitted to the different pretreatments.

As in the case of sugarcane bagasse and the other grasses samples, the spectra of *E. grandis x urophylla* samples pretreated with hot water (Figure 6a) or sodium bisulfite (Figure 6b) were very similar to that of the untreated sample, which confirms that these pretreatments are inefficient in the removal of hemicelluloses and lignin. Moreover, in samples pretreated with sulfuric acid (Figure 6c), the hemicellulose content was only decreased in response to pretreatment temperatures above 130°C, which was again very similar to the response of the novel grasses samples. However, sodium hydroxide was effective in the removal of hemicellulose at all pretreatment temperatures (Figure 6d), whereas lignin content was significantly reduced only in samples pretreated at 180°C. By contrast, pretreatment of sugarcane bagasse and the other investigated grasses at low temperatures appeared to be sufficient to reduce lignin content. Finally,

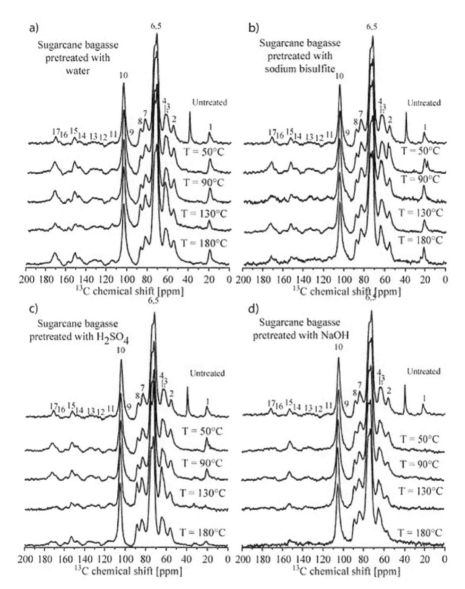

FIGURE 5: CPMASTOSS spectra of the solid fractions of sugarcane bagasse sample submitted to the different pretreatments.(a) hot water; (b) sodium bisulfite; (c) sulfuric acid and (d) sodium hydroxide pretreatments, respectively. Lines 3 and 7: C6 and C4 carbons from amorphous cellulose [38-42]; lines 4 and 8: C6 and C4 carbons [35-37]; lines 2, 11, 12, 13, 14: and 15: lignin carbons [37,43]; lines, 1, 3, 6, 7, 9 and 17: hemicelluloses carbons [36,44]; the unmarked line at 39 ppm is due to ash from biomass burned.

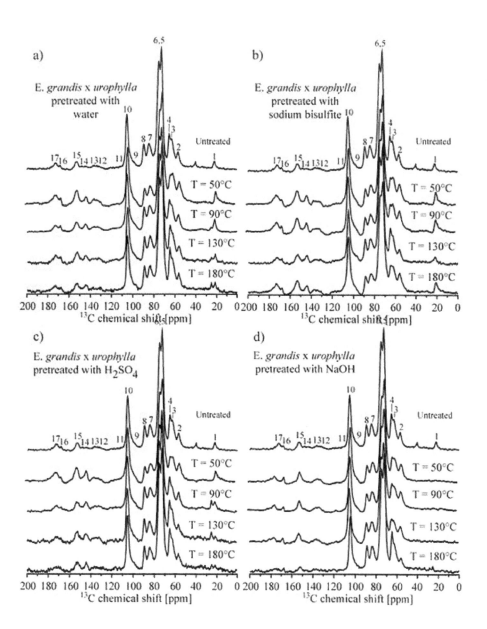

FIGURE 6: CPMASTOSS spectra of the solid fractions of E.grandis x urophylla barks samples submitted to the different pretreatments.(a) hot water; (b) sodium bisulfite; (c) sulfuric acid and (d) sodium hydroxide pretreatments, respectively.

the spectra of the sample pretreated with sodium hydroxide at 180°C also suggests that there was a reduction in the amorphous cellulose content in this sample after alkaline pretreatment. This should be compared to previously published results, using constant pretreatment temperature (120°), where higher sodium hydroxide concentrations (2% or 4%) and longer pretreatment times were required to remove the lignin fraction from these bark samples efficiently [15].

In summary, the NMR results indicated that, among the considered pretreatments, sulfuric acid was most effective in the removal of hemicellulose but sodium hydroxide was most efficient in the removal of hemicellulose together with a reduction in lignin content in both grasses and eucalyptus bark biomasses. However, the pretreatment temperature was also an important parameter and the use of higher temperatures promoted the removal of amorphous cellulose. In this sense, the results point to the intrinsic advantages of grass samples, which require lower pretreatment temperatures than eucalyptus barks.

2.2.5 SOLUBLE FRACTION ANALYSIS: MONOSACCHARIDE AND FURALDEHYDE CONTENT

To evaluate the generation of inhibitors and potential valuable products in the soluble phase of the protocol, a profile of compounds moved by the pretreatment solution was determined. The monosaccharide composition of the soluble fraction from hot water, sulfuric acid, sodium hydroxide and sodium bisulfite pretreatments at increasing temperatures, ranging from 50°C to 180°C (Figure 7), was studied. The potential formation of 2-furaldehyde and 5-hydroxymethyl-furaldehyde as a result of sulfuric acid pretreatment was also investigated in all six feedstocks (Figure 8).

For the pretreatments conducted at 50°C, glucose was the main monosaccharide in the soluble fraction from most of the biomasses and was detected together with xylose and other hemicellulose sugars (Figure 7). It can be related to an easier solubilization of glucose from hemicellulose, as well as the removal of the amorphous cellulose fraction. This enrichment in glucose was particularly evident in hot water, acid and sulfite pretreatments. In the soluble fraction from sodium hydroxide pretreatment, the

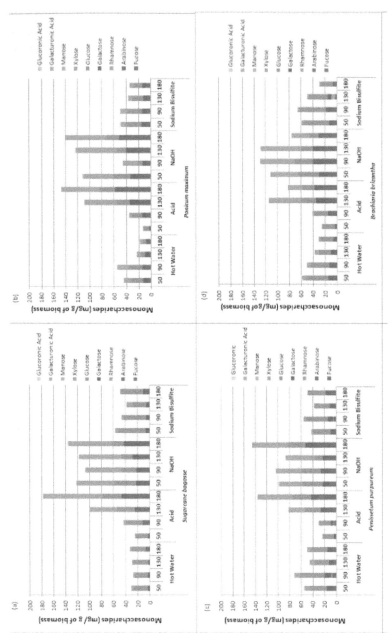

FIGURE 7: Monosaccharide composition in the liquor fraction from different pretreatments. (a) Sugarcane bagasse; (b)*Panicum maximum*; (c)*Pennisetum purpureum*; (d)*Brachiaria brizantha*; (e)*Eucalyptus grandis* bark; (f)*E. grandis x urophylla* bark. Pretreatment types and temperatures are indicated.

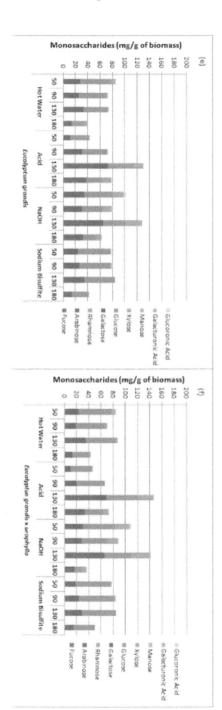

FIGURE 7: CONTINUED.

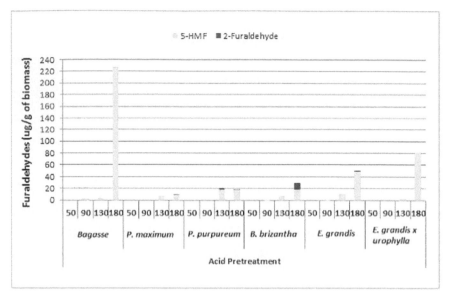

FIGURE 8: 2-furaldehyde and 5-hydroxymethyl-furfural content in the liquor fraction from acid pretreatment at increasing temperatures, ranging from 50°C to 180°C.

xylose amount was higher than glucose for all grasses even at 50°C, while for the bark samples the opposite was observed. This difference is associated with the efficient removal of the hemicellulose fraction by alkaline pretreatment, even at lower temperature, and the different composition of hemicelluloses in eucalyptus bark, which has a lower content of xylans.

With increasing temperatures, a gradual increase of xylose, arabinose, galactose and other monosaccharides was also observed for all pretreatments, indicating an efficient removal of the hemicellulose fraction. However, acid and alkaline pretreatments indicated a higher content of monosaccharides in the soluble fraction for all biomasses.

At higher temperature (180°C), a decrease of glucose content for all biomasses, in spite of xylose increase, became evident, most notably with acid pretreatment. The fall in glucose observed at higher temperatures can be explained by the formation of inhibitors, as shown in Figure 8. The highest 5- hydroxymethyl-furfural content was found for all biomasses pretreated at 180°C using sulfuric acid. However, lower amounts could be observed at 90°C or higher. Acidic conditions lead to a rapid decay of glucose into 5-hydroxymethyl-furfural by dehydration [45]. Sugarcane bagasse and bark were more susceptible to cellulose dehydration, whereas the perennial grasses showed levels below 20 µg of hydroxymethyl-furfural per gram of biomass. C5 conversion into 2-furaldehyde was found mainly in the soluble fraction from perennial grasses, with *B. brizantha* being most prone to the production of 2-furaldehyde under acid treatment.

2.2.6 MORPHOLOGICAL CHANGES PRODUCED BY PRETREATMENTS

To evaluate the effect of pretreatments on the morphology of different biomasses to improve enzymatic digestibility, we used scanning electron microscopy to investigate the morphological changes produced by sodium hydroxide pretreatment at 130°C. This pretreatment results in significant lignin and hemicellulose removal and, consequently, a higher cellulose enrichment, without the production of high levels of inhibitor.

Figure 9 shows the effects of different pretreatments on sugarcane bagasse, compared to raw material. A sample obtained from hot water pretreatment (Figure 9b) showed a similar surface to that obtained for raw bagasse (Figure 9a), where there was a continuous covering layer (possibly formed by lignin and hemicellulose). After acid pretreatment (Figure 9c), cellulose bundles were more evident, with less cohesion between them. This can be associated with the high level of hemicellulose removal, thereby enabling enzyme access to the cellulose fiber. A continuous layer over the cellulose bundles surface was also observed after sodium bisulfite treatment, but in this case some parts of the bundles were already evident, as shown in Figure 9d. Furthermore, it was possible to observe some residues over the surface, which could be associated with lignin modification and precipitation.

Among the pretreatments described here, the largest morphological changes were produced by sodium hydroxide. Figure 10 shows the effects of sodium hydroxide on sugarcane bagasse at 130°C, demonstrating the removal of the covering layer, mainly lignin (as determined by chemical composition), and a consequent loss of biomass structure, with separation

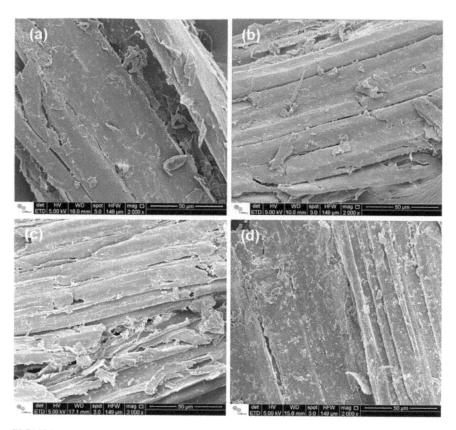

FIGURE 9: Scanning electron microscopy images of sugarcane bagasse before and after different pretreatments at 130°C. (a) Raw sugarcane bagasse (no pretreatment); (b) sugarcane bagasse pretreated with hot water; (c) sugarcane bagasse after sulfuric acid pretreatment and (d) sugarcane bagasse obtained after sodium bisulfite pretreatment.

of fiber bundles (Figure 10a). Lignin precipitation was also observed on the surface of fibers. At higher magnification, the presence of microfibrillated cellulose on the surface of samples could be observed (Figure 10b). Recently, such cellulose particles have been the focus of an exponentially increasing number of works, mainly interested in their structure and their potential to act as fillers to improve mechanical and barrier properties of biocomposites. Cellulose nanofillers are mainly native cellulose (cellulose I), extracted by traditional bleaching treatments of lignocellulosic

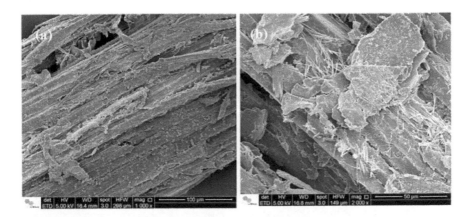

FIGURE 10: Scanning electron microscopy images of sugarcane bagasse submitted to sodium hydroxide pretreatment. (a) General view of sugarcane surface after alkali pretreatment and (b) higher magnification of cellulose whiskers on biomass surface.

fibers. However, the extraction conditions (time, temperature, chemical concentration) are fundamental to the efficient extraction of cellulose nanoparticles with the required characteristics [46]. These microfibrillated celluloses were not observed for sugarcane bagasse in a previous pretreatment condition using the same 1% sodium hydroxide concentration when preceded by a sulfuric acid pretreatment step at 120°C and a residence time for the alkaline step of 1 h [37]. In the present paper, however, the residence time was 40 min and treatment temperature was 130°C, without the acid step.

The effects of sodium hydroxide pretreatment at 130°C were also studied in the other biomass samples (Figure 11). Scanning electron microscopy images of the perennial grasses also revealed longer and isolated fibers of crystalline cellulose, compared to those of sugarcane bagasse (Figure 11a-d). This indicates their potential for the generation of natural fillers after efficient enzymatic hydrolysis, when the crystalline cellulose can be easily accessed and cleaved by cellulases. The surface images also revealed a notable effect of sodium hydroxide pretreatment at 130°C on eucalyptus bark surface, mainly due to the removal of lignin and its precipitation. However, microfibrillated cellulose was not detected on the

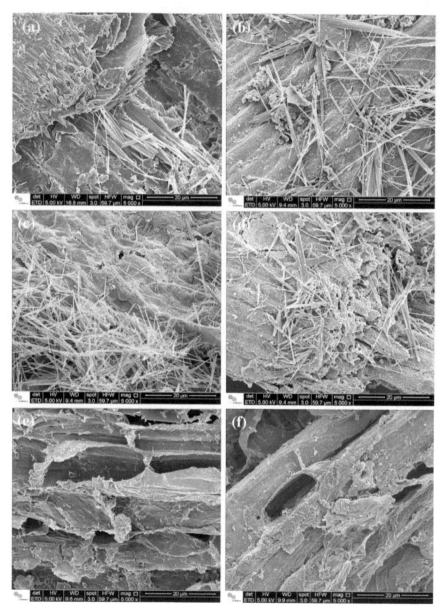

FIGURE 11: Scanning electron microscopy images of different biomasses pretreated with sodium hydroxide at 130°C, revealing the production on cellulose whiskers. (a) sugarcane bagasse; (b)*Panicum maximum*; (c)*Pennisetum purpureum*; (d)*Brachiaria brizantha*; (e) *Eucalyptus grandis* bark; and (f)*E. grandis x urophylla* bark.

barks surface, suggesting the need for more severe pretreatment conditions to obtain pure cellulose fibers.

2.2.7 CHANGES IN ENZYMATIC SACCHARIFICATION

Saccharification screening was performed to verify the effect of pretreatments on the saccharification potential of different biomasses. Results of this analysis indicated that sulfuric acid and sodium hydroxide greatly improved the sugar release from sugarcane bagasse and the three perennial grasses, whilst for the eucalyptus bark samples, sodium hydroxide pretreatment was significantly better (Figure 12). This differential effect could be related to the different hemicelluloses and different composition of lignin in eucalyptus bark.

The amounts of sugar released by the six feedstocks pretreated with hot water and sodium bisulfite were very similar to that of the control, and for all biomasses there was only a discrete effect of increased temperature, which correlates well with the results of the chemical analysis. However, the increase of pretreatment temperature significantly affected the enzymatic digestibility of sugarcane bagasse and grasses submitted to acid and sodium hydroxide. A gradual increase of sugar release was observed up to 130°C, followed by a decrease at 180°C. For the eucalyptus barks, the temperature effect was also very discrete, and even at the lower temperatures used, the glucose amount released was relatively high in both feedstocks.

2.3 CONCLUSION

The biomass feedstocks investigated in this work illustrate potential as a source of carbohydrates for bioethanol production. These feedstocks can be sustainably grown and applied to local production during the interseason, when no sugarcane bagasse is produced. Alkaline pretreatment at 130°C led to higher saccharification yields for the grass feedstocks, showing quite similar amounts of reducing sugars released for the three grasses and sugarcane bagasse. The alkaline treatment also resulted in

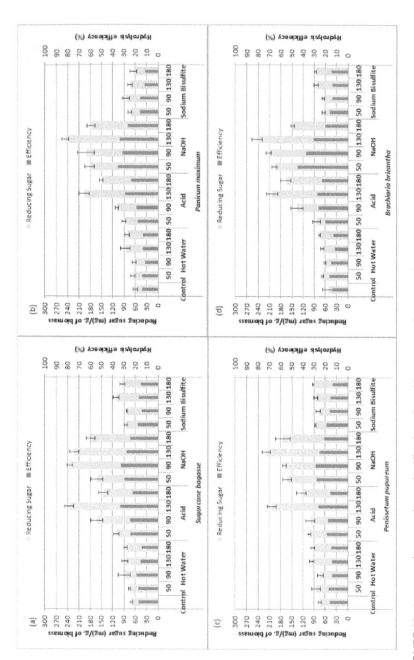

FIGURE 12: Automated enzymatic saccharification of raw biomasses and pretreated samples. (a) Sugarcane bagasse; (b)*Panicum maximum;* (c)*Pennisetum purpureum;* (d)*Brachiaria brizantha;* (e)*Eucalyptus grandis* bark; (f)*E. grandis x urophylla* bark. Pretreatment types and temperatures are shown.

higher glucose release for eucalyptus bark, even at the lower temperatures used. The relatively higher sugar yields obtained from sugarcane bagasse and the grasses, when compared to eucalyptus bark, can be explained by the morphological and chemical changes occurring during pretreatment. Chromatographic analysis indicated a higher cellulose enrichment in the grasses after sodium hydroxide pretreatment, and morphological analyses by scanning electron microscopy illustrated the effects of pretreatment on biomass structure, specifically the removal of lignin and the production of microfibrillated cellulose in grass samples, which justifies the documented improvement in enzymatic digestibility.

2.4 MATERIAL AND METHODS

2.4.1 PLANT MATERIAL

Sugarcane bagasse obtained after the industrial process for juice extraction for ethanol generation was kindly provided by the Cosan Group (Ibaté, SP, Brazil). Bark residues from mechanized harvesting of two commercial Eucalyptus clones (*E. grandis* and the hybrid *E. grandis x urophylla*) were provided by the Suzano Pulp and Paper Company (Itapetininga, SP, Brazil). Grass biomasses (*Pennisetum purpureum, B. brizantha* and *Panicum maximum*) grown for 180 days were supplied by Embrapa (Empresa Brasileira de Pesquisa Agropecuária). All the biomasses were dried in a convection oven at 60°C for 24 h and then ground to a fine powder in a ball mill (TissueLyser II, Qiagen (Hilden, Germany) for 2 min at 25 Hz.

2.4.2 DETERMINATION OF SILICON LEVEL BY X-RAY FLUORESCENCE SPECTROMETRY

XRF measurements were performed as previously described [47], using a commercial Portable XRF instrument (Niton XL3t900 Analyzer, Thermo Scientific, Hemel Hempstead, Hertfordshire, UK) equipped with an X-ray tube and a silicon drift detector. All measurements were carried out in a helium atmosphere with a helium flow rate of 70 centiliters min-1, and the samples were exposed to X-rays for 30 s (instrument settings: main range: 5 s, 6.2 kV, 100 uA, filter blank (no physical filter in the primary beam); low range: 25 s, 6.2 kV, 100 uA, filter blank). XRF experiments were performed on the raw powder biomasses, previously dried at 60°C for 24 h, and silica powder (Fisher Scientific, Loughborough, UK; product number S/0680/53) was used as standard to the calibration curve generation. To obtain a repeatable photon flux from the sample to the XRF detector, sample pellets were prepared using approximately 0.7 g of ground plant material, submitted to 11 tons for 2 s using a manual hydraulic press (Specac, Orpington, UK). Cylindrical pellets of around 5 mm thickness and 12 mm diameter were obtained and measured on both surface sides.

2.4.3 IMMUNOLABELING OF HEMICELLULOSIC POLYSACCHARIDES

The hemicelluloses fraction from biomasses (30 mg, excluding the soluble fraction) was obtained by extraction with 4 mL of 4 M sodium hydroxide/1% sodium borohydride for 1 h at room temperature. The extraction procedure was repeated twice and the supernatants were combined and neutralized with acetic acid. The polysaccharides from the hemicellulose fraction were precipitated with ethanol and solubilized in water (10 mL). Following protocol optimization, specific polysaccharide dilutions were used for each antibody: LM10 (40× dilution for all biomasses); LM11 (40× dilution for all biomasses); and LM21 (20× dilution for all biomasses). Monoclonal antibodies were obtained as hybridoma cell culture supernatants from Plant Probe Laboratories (Leeds, UK) and diluted 25-fold for all assays. Selected antibodies were: LM10 and LM11, which recognize a xylan and arabinoxylan epitope respectively; and LM21, which primarily recognizes mannan, galactomannan and glucomannan polysaccharides [30-32]. Stock solutions (10 μg/mL in deionized water) of beech wood xylan (x4252; Sigma-Aldrich, Seelze, Germany), wheat arabinoxylan (70502a; Megazyme, Bray, County Wicklow, Ireland) and Konjac glucomannan (50601; Megazyme) were used as positive control for LM10, LM11 and LM21, respectively. The immunolabeling assays were performed as previously described and the degree of antibody binding was measured at 620 nm after a reaction time of 40 min to enable color development.

2.4.4 PRETREATMENT

Plant materials (400 mg) were pretreated in small-scale bombs, using 16 mL of pretreatment solution (hot water, 0.1 M sulfuric acid, 0.25 M sodium hydroxide and 3% (w/v) sodium bisulfite). Pretreatment was performed for 40 min in the oven at different temperatures (50°C, 90°C, 130°C and 180°C). After cooling to the room temperature, samples were transferred to 50 mL centrifuge tubes and centrifuged at 4,000 g for 15 min. The liquor fraction was recovered into a new tube and stored for further analysis. The solid fraction was resuspended in 5 mL of ethanol, then homogenized and centrifuged at 4,000 g for 15 min. An ethanol washing procedure was conducted in triplicate before the biomass was dried at room temperature.

2.4.5 CHEMICAL ANALYSIS

2.4.5.1 SOLUBLE EXTRACTION

The soluble content was determined by sequential extraction with different organic solvents. Powdered plant material (200 mg) was weighed into a centrifuge tube and extracted with phenol (5 mL), under constant vortex agitation for 1 min. Samples were centrifuged for 20 min at 10,000 g, after which the supernatant was discarded. The pellet was resuspended in 10 mL of a chloroform:methanol (2:1 v/v) solution and homogenized vigorously for 2 min. The sample was centrifuged as previously and the pellet was washed with ethanol (2 mL) twice more. After washing, 5 mL of 90% dimethyl sulfoxide was added to each sample, which was then and left rocking overnight at room temperature. The dimethyl sulfoxide supernatant obtained by centrifugation was removed and the solid fraction was washed three times with ethanol absolute. Finally, samples were dried in a vacuum dryer and their dry weight was recorded. The difference between the initial and final weights was used to determine the soluble fraction (%).

2.4.5.2 DIGESTION OF NON-CRYSTALLINE POLYSACCHARIDES FOR MONOSACCHARIDE ANALYSIS

Pretreated samples and soluble extracted biomass samples (10 mg) were weighed in a 2 mL capped tube and incubated in 0.5 mL of 2 M trifluoroacetic acid (TFA) for 4 h at 100°C in an argon atmosphere. Samples were homogenized every hour, then the TFA was evaporated in a centrifugal evaporator at 45°C. Samples were resuspended in 0.5 mL of Miliq water under vigorously agitation and centrifuged at 10,000 g for 10 min. The supernatant was recovered into a new tube without disturbing the pellet and the washing procedure repeated once. The water soluble monosaccharides in the supernatant (hemicellulose fraction) were vacuum-dried and stored at room temperature for further chromatographic analysis. The remaining solid fraction was used to determine the crystalline cellulose content.

2.4.5.3 CRYSTALLINE CELLULOSE CONTENT

The residual solid fraction obtained after TFA extraction (as described above) was initially submitted to hydrolysis using 1 mL of Updegraff reagent [48]. Samples were vortexed then incubated at 100°C for 30 min. After cooling to room temperature, samples were centrifuged and the supernatant was carefully discarded without disturbing the remaining crystalline cellulose fraction. The pellet was washed with 1.5 mL of water and centrifuged. Three additional washes were performed using 1.5 mL of acetone before the samples were dried at room temperature. The hydrolysis of crystalline cellulose was performed at room temperature for 30 min, using 175 μL of 72% (v/v) sulfuric acid. The reaction was stopped by the addition of 825 μL of water and then the sample was vortexed. Finally the sample was centrifuged before the glucose content was determined by the anthrone method [49].

2.4.5.4 ACETYL BROMIDE SOLUBLE LIGNIN

The total lignin content of soluble extracted plant material and samples obtained from each pretreatment condition was determined using the acetyl bromide method [50]. Powder biomass (3.5 mg) was weighed into a 2 mL cap tube and 250 µL of freshly prepared acetyl bromide solution (25% v/v acetyl bromide/glacial acetic acid) was added. Samples were incubated at 50°C for 3 h, with periodical agitation. After cooling to room temperature, the hydrolysate was transferred to a 5 mL volumetric flask. One mL of 2 M sodium hydroxide was added to the 2 mL tube to generate the acetyl bromide excess before transfer to a volumetric flask. Next, 175 µL of hydroxylamine-hydrochloric acid was added to each sample, which were then vortexed vigorously. Finally, the volume was adjusted to 5 mL with glacial acetic acid and the absorbance was measured at 280 nm. The acetyl bromide soluble lignin (%) was determined using the extinction coefficient for grasses (17.75).

2.4.5.5 MONOSACCHARIDE ANALYSIS

Monosaccharide analysis was performed by high performance anion-exchange chromatography (Dionex IC 2500; Thermo Scientific, Camberley, Surrey, UK) on a Dionex Carbopac PA-10 column with integrated amperometry detection [51]. The separated monosaccharides were quantified using external calibration with an equimolar mixture of nine monosaccharide standards (arabinose, fucose, galactose, galacturonic acid, glucose, glucuronic acid, mannose, rhamnose and xylose), which were subjected to TFA hydrolysis in parallel with the samples.

2.4.6 SUGAR CONTENT IN PRETREATMENT LIQUORS

Monosaccharide composition in the liquor fraction obtained from each pretreatment condition was also determined by ion chromatography. Acidic liquor samples were initially neutralized with barium hydroxide solution (150 mM), followed by barium carbonate powder. Alkaline samples were neutralized with 2 M hydrochloric acid. All the samples were adjusted to the same final volume and centrifuged at 4,000 g to warranting precipitate removal. Sulfite liquor samples were filtered using an ion exchange column (onGuard II Ba Cartridge, Dionex) to eliminate the residual ions from the samples.

The neutralized liquor fraction (1 mL) was transferred to a microcentrifuge tube and vacuum. Hydrolysis using 500 μL of 2 M TFA was carried out at 100°C for 4 h. After cooling to room temperature, samples were dried and twice washed with 200 μL of isopropanol. Monosaccharide content was determined as previously described.

2.4.7 ANALYSIS OF FURFURAL AND 5-HYDROXYMETHYL-FURFURAL CONTENT

Liquor fractions obtained from each pretreatment condition were neutralized and subjected to chromatography using a Luna® 5 μm C18(2) 100 Å LC Column 150 × 4.6 mm, together with a C18 4 × 2.0 mm ID guard column (both from Phenomenex, Cheshire, UK) to verify furfural and 5-hydroxymethyl-furfural content. Analyses were carried out using a Surveyor HPLC (Thermo Scientific, Hemel Hempstead, UK), with an elution system of acetonitrile by reversed-phase in an isocratic gradient (5% acetonitrile and 95% deionized water) at 1 mL/min. The eluted furfuraldehydes were detected by UV absorbance at 284 nm using a Finnigan Surveyor PDA Plus detector (Thermo Scientific; Hemel Hempstead, UK). The furfurals were quantified by interpolation of a calibration curve within the range of 0.005 μg/mL to 50 μg/mL in water.

2.4.8 SOLID-STATE NUCLEAR MAGNETIC RESONANCE

NMR experiments were performed in a Varian INOVA spectrometer (Varian; Palo Alto, California, USA) operating at 13C and 1H frequencies of 100.5 and 400.0 MHz, respectively. A Varian 5-mm magic angle spinning double-resonance probe head was used. The spinning frequency of 4.5 kHz was controlled by a Varian pneumatic system ensuring a rotation stability of approximately 2 Hz. CPMASTOSS was used for 13C excitation in all experiments. Typical $\pi/2$ pulse lengths of 3.5 μs (13C) and 4.5 μs (1H), continuous wave proton decoupling with field strength of 60 kHz, cross-polarization time of 1 ms and recycle delays of 2 s were used .

2.4.9 SCANNING ELECTRON MICROSCOPY

Surface images of the grasses and both eucalyptus barks after variable pretreatment conditions were obtained by scanning electron microscopy and compared with the raw material. Milled samples were critical point dried before coating with gold in a Balzers SCD 050 sputter coater (BAL-TEC AG, Balzers, Liechtenstein). Samples were viewed using a scanning electron microscope model Quanta 650-FEG (FEI, Hillsboro, Oregon, USA) from the National Laboratory of Nanotechnology (LNNano-CNPEM) in Campinas, SP, Brazil. A large number of images were obtained from different areas of the samples (at least 20 images per sample) to confirm the reproducibility of results.

2.4.10 AUTOMATED ENZYMATIC SACCHARIFICATION ANALYSIS

Automated saccharification assays were performed as previously described [52]. Powder pretreated samples and soluble extracted biomasses were dispensed into 96-well plates using a custom-made robotic platform (Labman Automation Ltd., Stokesley, UK) and the standard target weight of material was 4 mg. Enzymatic hydrolysis and sugar determination were performed automatically using a robotic platform (Tecan Evo 200; Tecan Group Ltd., Männedorf, Switzerland). The hydrolysis was carried out in a monitored shaking incubator (Tecan Group Ltd.) using an enzyme cocktail with a 4:1 ratio of Celluclast and Novozyme 188 (cellobiase from Aspergillus niger; both from Novozymes, Bagsvaerd, Denmark) at 30°C for 8 h. The saccharification of the powdered biomass was performed in a total volume of 600 μl for 8 h, with an enzyme loading of 8 FPU/g of biomass. Automated determination of released reducing sugar after hydrolysis was performed using 3-methyl-2-benzothiazolinone hydrozone, as previously described and established [52,53].

REFERENCES

1. Farrell AE, Plevin RJ, Turner BT, Jones AD, O'Hare M, Kammen DM: Ethanol can contribute to energy and environmental goals. Science 2006, 311:506-508.
2. Havlík P, Schneider UA, Schmid E, Böttcher H, Fritz S, Skalský R, Aoki K, Cara SD, Kindermann G, Kraxner F, Leduc S, McCallum I, Mosnier A, Sauer T, Obersteiner M: Global land-use implications of first and second generation biofuel targets. Energy Policy 2011, 39:5690-5702.
3. Rosegrant MW, Zhu T, Msangi S, Sulser T: Global scenarios for biofuels: impacts and implications. App Econ Perspect Policy 2008, 30:495-505.
4. Lora ES, Andrade RV: Biomass as energy source in Brazil. Renew Sust Energ Rev 2009, 13:777-788.
5. La Rovere EL, Pereira AS, Simões AF: Biofuels and sustainable energy development in Brazil. World Dev 2011, 39:1026-1036.
6. Hill J: Environmental costs and benefits of transportation biofuel production from food- and lignocellulose-based energy crops: a review. In Sustainable Agriculture. Netherlands: Springer; 2009:125-139.
7. Granda CB, Zhu L, Holtzapple MT: Sustainable liquid biofuels and their environmental impact. Environ Prog 2007, 26:233-250.
8. Seabra JEA, Macedo IC: Comparative analysis for power generation and ethanol production from sugarcane residual biomass in Brazil. Energ Policy 2011, 39:421-428.
9. Conab: Acompanhamento da safra brasileira de cana-de-açúcar. Brasília-DF: Companhia Nacional de Abastecimento; 2012. http://www.conab.gov.br/OlalaCMS/uploads/arquivos/12_12_12_10_34_43_boletim_cana_portugues_12_2012.pdf
10. Yu Q, Zhuang X, Yuan Z, Wang Q, Qi W, Wang W, Zhang Y, Xu J, Xu H: Two-step liquid hot water pretreatment of Eucalyptus grandis to enhance sugar recovery and enzymatic digestibility of cellulose. Bioresour Technol 2010, 101:4895-4899.
11. Flynn B: Eucalyptus: having an impact on the global solid wood industry. 2010. http://www.wri-ltd.com/marketPDFs/Eucalyptus.pdf
12. Zhu JY, Pan XJ: Woody biomass pretreatment for cellulosic ethanol production: technology and energy consumption evaluation. Bioresour Technol 2010, 101:4992-5002.
13. Perlack RD, Wright LL, Turhollow A, Graham RL, Stokes B, Erbach DC: Biomass as feedstock for a bioenergy and bioproducts industry: the technical feasibility of a billion-ton annual supply. Oak Ridge: Oar Ridge National Laboratory; 2005. http://www.ornl.gov/~webworks/cppr/y2001/rpt/123021.pdf
14. Foelkel C: Eucalyptus Online Book. 2010. http://www.eucalyptus.com.br

15. Lima MA, Lavorente GB, da Silva H, Bragatto J, Rezende CA, Bernardinelli OD, de Azevedo ER, Gomez LD, McQueen-Mason S, Labate CA, Polikarpov I: Effects of pretreatment on morphology, chemical composition and enzymatic digestibility of eucalyptus bark: a potentially valuable source of fermentable sugars for biofuel production - part 1. Biotechnol Biofuels 2013, 6:75.

16. Peres AR, Vazquez GH, Cardoso RD: Physiological potential of *Brachiaria brizantha* cv. Marandu seeds kept in contact with phosphatic fertilizers. Revista Brasileira de Sementes 2012, 34:424-432.

17. Anualpec: Anuário da pecuária brasileira. São Paulo: Instituto FNP; 2011:378.

18. Karia CT, Duarte JB, Araújo ACG: Desenvolvimento de cultivares do gênero Brachiaria (trin.) Griseb. no Brasil. Planaltina, DF: Emprapa; 2006:58.

19. Dowe N, Mcmillan J: SSF experimental protocols - lignocellulosic biomass hydrolysis and fermentation. Golden, CO: National Renewable Energy Laboratory (NREL); 2008. Technical Report TP-510-42630

20. Lima Filho OF, Grothge-Lima MT, Tsai SM: Supressão de patógenos em solos induzida por agentes abióticos: o caso do silício. Informações Agronômicas 1999, 87:8-12.

21. Cennatek: Feasibility of improving biomass combustion through extraction of nutrients. 2011. http://www.ofa.on.ca/uploads/userfiles/files/cennatek%20ofa%20report-feasibility%20of%20improving%20biomass%20combustion%20through%20extraction%20of%20nutrients.pdf

22. Demirbas A: Potential applications of renewable energy sources, biomass combustion problems in boiler power systems and combustion related environmental issues. Prog Energ Combust Sci 2005, 31:171-192.

23. Khan AA, de Jong W, Jansens PJ, Spliethoff H: Biomass combustion in fluidized bed boilers: potential problems and remedies. Fuel Process Technol 2009, 90:21-50.

24. Emmel A, Mathias AL, Wypych F, Ramos LP: Fractionation of Eucalyptus grandis chips by dilute acid-catalysed steam explosion. Bioresour Technol 2003, 86:105-115.

25. Foelkel C: Casca da árvore do eucalipto. Eucalyptus Online Book 2006. http://www.eucalyptus.com.br

26. Lu H, Liu K: Phytoliths of common grasses in the coastal environments of southeastern USA. Estuar Coast Shelf Sci 2003, 58:587-600.

27. Nlewem KC, Thrash ME Jr: Comparison of different pretreatment methods based on residual lignin effect on the enzymatic hydrolysis of switchgrass. Bioresour Technol 2010, 101:5426-5430.

28. Goldemberg J: The Brazilian biofuels industry. Biotechnol Biofuels 2008, 1:6.

29. Bragatto J: Avaliação do potencial da casca de Eucalyptus spp. para produção de bioetanol. PhD thesis. University of São Paulo; 2010:154.

30. McCartney L, Marcus SE, Knox JP: Monoclonal antibodies to plant cell wall xylans and arabinoxylans. J Histochem Cytochem 2005, 53:543-546.

31. Marcus S, Verhertbruggen Y, Herve C, Ordaz-Ortiz J, Farkas V, Pedersen H, Willats W, Knox JP: Pectic homogalacturonan masks abundant sets of xyloglucan epitopes in plant cell walls. BMC Plant Biol 2008, 8:60.

32. Pattathil S, Avci U, Baldwin D, Swennes AG, McGill JA, Popper Z, Bootten T, Albert A, Davis RH, Chennareddy C, Dong R, O'Shea B, Rossi R, Leoff C, Freshour

G, Narra R, O'Neil M, York WS, Hahn MG: A comprehensive toolkit of plant cell wall glycan-directed monoclonal antibodies. Plant Physiol 2010, 153(2):514-525.

33. Wyman CE, Dale BE, Elander RT, Holtzapple M, Ladisch MR, Lee YY: Coordinated development of leading biomass pretreatment technologies. Bioresour Technol 2005, 96:1959-1966.

34. Tomás-Pejó E, Alvira P, Ballesteros M, Negro MJ: Pretreatment technologies for lignocellulose-to-bioethanol conversion. In Biofuels. Amsterdam: Elsevier, Academic Press; 2011:149-176.

35. Wickholm K, Larsson PT, Iversen T: Assignment of non-crystalline forms in cellulose I by CP/MAS C-13 NMR spectroscopy. Carbohydrate Res 1998, 312:123-129.

36. Templeton DW, Scarlata CJ, Sluiter JB, Wolfrum EJ: Compositional analysis of lignocellulosic feedstocks. 2. Method uncertainties. J Agric Food Chem 2010, 58:9054-9062.

37. Rezende CA, Lima MA, Maziero P, deAzevedo ER, Garcia W, Polikarpov I: Chemical and morphological characterization of sugarcane bagasse submitted to a delignification process for enhanced enzymatic digestibility. Biotechnol Biofuels 2011, 4:54.

38. Focher B, Marzetti A, Cattaneo M, Beltrame PL, Carniti P: Effects of structural features of cotton cellulose on enzymatic hydrolysis. J Appl Polym Sci 1981, 26:1989-1999.

39. Hallac BB, Sannigrahi P, Pu Y, Ray M, Murphy RJ, Ragauskas AJ: Biomass characterization of Buddleja davidii: a potential feedstock for biofuel production. J Agric Food Chem 2009, 57:1275-1281.

40. El Hage R, Brosse N, Sannigrahi P, Ragauskas A: Effects of process severity on the chemical structure of Miscanthus ethanol organosolv lignin. Polym Degrad Stabil 2010, 95:997-1003.

41. Sannigrahi P, Miller SJ, Ragauskas AJ: Effects of organosolv pretreatment and enzymatic hydrolysis on cellulose structure and crystallinity in Loblolly pine. Carbohydr Res 2010, 345:965-970.

42. Foston MB, Hubbell CA, Ragauskas AJ: Cellulose isolation methodology for NMR analysis of cellulose ultrastructure. Materials 2011, 4:1985-2002.

43. Martínez AT, González AE, Valmaseda M, Dale BE, Lambregts MJ, Solid-State HJF: Solid-state NMR studies of lignin and plant polysaccharide degradation by fungi. Holzforschung 1991, 45:49-54.

44. Wickholm K, Larsson PT, Iversen T: Assignment of non-crystalline forms in cellulose I by CP/MAS 13C NMR spectroscopy. Carbohydr Res 1998, 312:123-129.

45. Binder JB, Raines RT: Fermentable sugars by chemical hydrolysis of biomass. Proc Natl Acad Sci USA 2010, 107:4516-4521.

46. Siqueira G, Bras J, Dufresne A: Cellulosic bionanocomposites: a review of preparation, properties and applications. Polymers 2010, 2:728-765.

47. Reidinger S, Ramsey M, Hartley SE: Rapid and accurate analyses of silicon and phosphorus in plants using a portable X-ray fluorescence spectrometer. New Phytol 2012, 195:699-706.

48. Updegraff DM: Semimicro determination of cellulose in biological materials. Anal Biochem 1969, 32:420-424.

49. Loewus FA: Improvement in anthrone method for determination of carbohydrates. Anal Chem 1952, 24:219-219.
50. Fukushima RS, Hatfield RD: Extraction and isolation of lignin for utilization as a standard to determine lignin concentration using the acetyl bromide spectrophotometric method. J Agric Food Chem 2001, 49:3133-3139.
51. Jones L, Milne JL, Ashford D, McQueen-Mason SJ: Cell wall arabinan is essential for guard cell function. Proc Natl Acad Sci 2003, 100:11783-11788.
52. Gomez L, Whitehead C, Barakate A, Halpin C, McQueen-Mason SJ: Automated saccharification assay for determination of digestibility in plant materials. Biotechnol Biofuels 2010, 3:23.
53. Anthon GE, Barrett DM: Determination of reducing sugars with 3-methyl-2-benzo-thiazolinonehydrazone. Anal Biochem 2002, 305:287-289.

PART II

CULTIVATION AND OPTIMIZATION PROCESSES

CHAPTER 3

TOWARDS THE PRODUCTION OF SECOND GENERATION ETHANOL FROM SUGARCANE BAGASSE IN BRAZIL

T. P. BASSO, T. O. BASSO, C. R. GALLO AND L. C. BASSO

3.1 INTRODUCTION

Brazil and the United States produce ethanol mainly from sugarcane and starch from corn and other grains, respectively, but neither resource are sufficient to make a major impact on world petroleum usage. The so-called first generation (1G) biofuel industry appears unsustainable in view of the potential stress that their production places on food commodities. On the other hand, second generation (2G) biofuels produced from cheaper and abundant plant biomass residues, has been viewed as one plausible solution to this problem [1]. Cellulose and hemicellulose fractions from lignocellulosic residues make up more than two-thirds of the typical biomass composition and their conversion into ethanol (or other chemicals) by an economical, environmental and feasible fermentation process would be possible due to the increasing power of modern biotechnology and (bio)-process engineering [2].

Brazil is the major sugar cane producer worldwide (ca. 600 million ton per year). After sugarcane milling for sucrose extraction, a lignocellulosic residue (sugarcane bagasse) is available at a proportion of ca. 125 kg of dried bagasse per ton of processed sugarcane. Therefore, sugarcane bagasse is a suitable feedstock for second generation ethanol coupled to the first generation plants already in operation, minimizing logistic and energetic costs.

State-owned energy group Petrobras is one of the Brazilian groups leading the development of second generation technologies, estimating that commercial production could begin by 2015. Other organizations making significant contributions to next generation biofuels in Brazil include the Brazilian Sugarcane Technology Centre (CTC), operating a pilot plant for the production of ethanol from bagasse in Piracicaba (Sao Paulo) [3]. Recently, GraalBio (Grupo Brasileiro Graal) has stated publically that will start production of bioethanol from sugarcane bagasse in one plant located at the Northeast region of the country.

Pretreatment, hydrolysis and fermentation can be performed by various approaches. According to a CTC protocol, the process of manufacturing ethanol from bagasse is divided into the following steps. First, the bagasse is pretreated via steam explosion (with or without a mild acid condition) to increase the enzyme accessibility to the cellulose and promoting the hemicellulose hydrolysis with a pentose stream. The lignin and cellulose solid fraction is subjected to cellulose hydrolysis, generating a hexose-rich stream (mainly composed of glucose, manose and galactose). The final solid residue (lignin and the remaining recalcitrant cellulose) is used for heating and steam generation. The hexose fraction is mixed with 1G cane molasses (as a source of minerals, vitamins and aminoacids) and fermented by regular *Saccharomyces cerevisiae* industrial strains (not genetically modified) using the same fermentation and distillation facilities of the Brazilian ethanol plants. The pentose fraction will be used as substrate for other biotechnological purposes, including ethanol fermentation.

Researchers are focusing on cutting the cost of the enzymes and the pretreatment process, as well as reducing energy input. Production of ethanol from sugarcane bagasse will have to compete with the use of bagasse for electricity cogeneration. Depending on the efficiency of the cogeneration plant, about half of the bagasse is required to produce captive energy in the form of steam and energy at the sugar and ethanol facility. It is estimated that the surplus bagasse could increase the Brazilian ethanol production by roughly 50% [3].

3.2 LIGNOCELLULOSIC RESIDUES

Lignocellulosic residues are composed by three main components: cellulose, hemicellulose and lignin. Cellulose and hemicellulose are polysaccharides composed by units of sugar molecules [4]. Sugarcane bagasse is composed of around 50% cellulose, 25% hemicellulose and 25% lignin. It has been proposed that because of its low ash content (around 2-3%), this product offers numerous advantages in comparison to other crop residues (such as rice straw and wheat straw) when used for bio-processessing purposes. Additionally, in comparison to other agricultural residues, bagasse is considered a richer solar energy reservoir due to its higher yields on mass/area of cultivation (about 80 t/ha in comparison to about 1, 2, and 20 t/ha for wheat, other grasses and trees, respectively) [5].

3.2.1 CELLULOSE

Cellulose is composed of microfibrils formed by glucose molecules linked by β-1,4, being each glucose molecule reversed in relation to each other. The union of microfibrils form a linear and semicrystalline structure. The linearity of the structure enables a strong bond between the microfibrils. The crystallinity confers resistance to hydrolysis due to absence of water in the structure and the strong bond between the glucose chains prevents hydrolases act on the links β -1,4 [1].

3.2.2 HEMICELLULOSE

Hemicellulose is a polysaccharide made of polymers formed by units of xylose, arabinose, galactose, manose and other sugars, that present cross-linking with glycans. Hemicellulose can bind to cellulose microfibrils by hydrogen bonds, forming a protection that prevents the contact between microfibrils to each other and yielding a cohesive network. Xyloglucan is the major hemicellulose in many primary cell walls. Nevertheless, in secondary cell wall, which predominate in the plant biomass, the hemicelluloses are typically more xylans and arabino-xylans. Typically, hemicellulose comprises between 20 to 50% of the lignocellulose polysaccharides, and therefore contributes significantly to the production of liquid biofuels [1]. Sugarcane bagasse contain approximately 25-30% hemicelluloses [13,14].

3.2.3 LIGNIN

Lignin is a phenolic polymer made of phenylpropanoid units [13], which has the function to seal the secondary cell wall of plants. Besides providing waterproofing and mechanical reinforcement to the cell wall, lignin forms a formidable barrier to microbial digestion. Lignin is undoubtedly the most important feature underlying plant biomass architecture. Sugarcane bagasse and leaves contain approximately 18-20% lignin [13]. The phenolic structure of this polymer confers a material that is highly resistant to enzymatic digestion. Its disruption represents the main target of pretreatments before enzymatic hydrolysis.

3.3 PRETREATMENT

The pretreatment process is performed in order to separate the carbohydrate fraction of bagasse and other residues from the lignin matrix. Another function is to minimize chemical destruction of the monomeric sugars [6]. During pretreatment the inner surface area of substrate particles is enlarged by partial solubilization of hemicellulose and lignin.. This is achieved by various physical and/or chemical methods [5]. However, it has been generally accepted that acid pretreatment is the method of choice in several model processes [7]. One of the most cost-effective pretreatments is the use of diluted acid (usually between 0.5 and 3% sulphuric acid) at moderate temperatures. Albeit lignin is not removed by this process, its disruption renders a significant increase in sugar yield when compared to other processes [1]. Regarding sugarcane bagasse several attempts have been made to optimize the release of the carbohydrate fraction from the lignin matrix, including dilute acid pretreatment, steam explosion, liquid hot water, alkali, peracetic acid and also the so called ammonia fiber expansion (AFEX).

3.4 HYDROLYSIS

Cellulose and hemicellulose fractions released from pretreatment has to be converted into glucose and other monomeric sugars. This can be achieved by both chemically or enzymatically oriented hydrolysis [7, 8].

3.4.1 CHEMICAL HYDROLYSIS

Whitin chemical hydrolysis, acid hydrolysis is the most used and it can be performed with several types of acids, including sulphurous, sulphuric, hydrochloric, hydrofluoric, phosphoric, nitric and formic acid. While processes involving concentrated acids are usually operated at low temperatures, the large amount of acids required may result in problems associated with equipment corrosion and energy demand for acid recovery. These processes typically involve the use of 60-90% concentrated sulfuric acid. The primary advantage of the concentrated acid process in realtion to diluted acid hydrolysis is the high sugar recovery efficiency, which can be on the order of 90% for both xylose and glucose. Concentrated acid hydrolysis disrupts the hydrogen bonds between cellulose chains, converting it into a completely amorphous state [8].

On the other hand, during dilute acid hydrolysis temperatures of 200–240°C at 1.5% acid concentrations are required to hydrolyze the crystalline cellulose. Besides that, pressures of 15 psi to 75 psi, and reaction time in the range of 30 min to 2 h are employed. During this conditions degradation of monomeric sugars into toxic compounds and other non-desired products are inevitable [9]. The main advantage of dilute acid hydrolysis in comparison to concentrated acid hydrolysis is the relatively low acid consumption. However, high temperatures required to achieve acceptable rates of conversion of cellulose to glucose results in equipment corrosion [4].

3.4.2 ENZYMATIC HYDROLYSIS

Differently from acid hydrolysis, biodegradation of sugarcane bagasse by cellulolytic enzymes can be performed at much lower temperatures (around 50°C or even lower). Moreover conversion of cellulose and hemicellulose polymers into their constituent sugars is very specific and toxic degradation products are unlikely to be formed. However, a pretreatment step is required for enzymatic hydrolysis, since the native cellulose structure is well protected by the matrix compound of hemicellulose and lignin [4].

Cellulase is the general term for the enzymatic complex able to degrade cellulose into glucose molecules. The mechanism action accepted for hydrolysis of cellulose are based on synergistic activity between endoglucanase (EC 3.2.1.4), exoglucanase (or cellobiohydrolase (EC 3.2.1.91)), and β-glucosidase (EC 3.2.1.21). The first enzyme cleaves the bounds β-1,4-glucosidic of cellulose chains to produce shorter cello-dextrins. Exoglucanase release cellobiose or glucose from cellulose and cello-dextrin chains and, finally β-glucosidases hydrolyze cellobiose to glucose. The intramolecular β-1,4-glucosidic linkages are cleaved by endoglucanases randomly. Endoglucanases and exoglucanases have different modes of action. While endoglucanase hydrolyze intramolecular cleavages, exoglucanases hydrolyze long chains from the ends. More specifically, exoglucanases or cellobiohydrolases have action on the reducing (CBH I) and non-reducing (CBH II) cellulose chain ends to liberate glucose and cellobiose. These enzymes acts on insoluble cellulose, then their activity are often measured using microcrystalline cellulose. Lastly, β-glucosidases or cellobioase hydrolyze cellobiose to glucose. They are important to the process of hydrolysis because they removed cellobiose to the aqueous phase that is an inhibitor to the action of endoglucanases and exoglucanases [10].

The multi-complex enzymatic cocktail known as cellulase and hemicellulase can be produced by a variety of saprophytic microorganisms.

Trichoderma and *Aspergillus* are the genera most used to produce cellulases. Among them, one of the most productive of biomass degrading enzymes is the filamentous fungus *Trichoderma reesei*. It cellulolytic arsenal is composed by a mixture of endoglucanases and exoglucanases that act synergistically to break down cellulose to cellobiose. Two β-glucosidases have been identified that are implicated in hydrolyzing cellobiose to glucose. An additional protein, swollenin, has been described that disrupts crystalline cellulose structures, presumably making polysaccharides more accessible to hydrolysis. The four most abundant components of *T. reesei* cellulase together constitute more than 50% of the protein produced by the cell under inducing conditions [9, 15]. Cellulases are essential for the biorefinery concept. In order to reduce the costs and increase production of commercial enzymes, the use of cheaper raw materials as substrate for enzyme production and focus on a product with a high stability and specific activity are mandatory. Apart from bioethanol, there are several applications to these enzymes, such as in textile, detergent, food, and in the pulp and paper industries.

3.5 ENZYME PRODUCTION USING SUGARCANE BAGASSE

Cultivation of microorganisms in agroindustrial residues (such as bagasse) aiming the production of enzymes can be divided into two types: processes based on liquid fermentation or submerged fermentation (SmF), and processes based on solid-state fermentation (SSF) [5]. In several SSF processes bagasse has been used as the solid substrate. In the majority of the processes bagasse has been used as the carbon (energy) source, but in some others it has been used as the solid inert support. Cellulases have been extensively studied in SSF using sugarcane bagasse. It has been reported the production of cellulases from different fungal strains [5].

Several processes have been reported for the production of enzymes using bagasse in SmF. One of the most widely studied aspects of bagasse application has been on cellulolytic enzyme production. Generally basidiomycetes have been employed for this purpose, in view of their high extracellular cellulase production. A recent example was the use of Trichoderma reesei QM-9414 for cellulase and biomass production from bagasse.

Additionally, white-rot fungi were successfully used for the degradation of long-fiber bagasse. Most of the strains caused an increase in the relative concentration of residual cellulose, indicating that hemicellulose was the preferred carbon source [5].

3.6 PROCESSES FOR ETHANOL PRODUCTION

Ethanol production from lignocellulosic residues is performed by fermentation of a mixture of sugars in the presence of inhibiting compounds, such as low molecular weight organic acids, furan derivatives, phenolics, and inorganic compounds released and formed during pretreatment and/ or hydrolysis of the raw material. Ethanol fermentation of pentose sugars (xylose, arabinose) constitutes a challenge for efficient ethanol production from these residues, because only a limited number of bacteria, yeasts, and fungi can convert pentose (xylose, arabinose), as wells as other monomers released from hemicelluloses (mannose, galactose) into ethanol with a satisfactory yield and productivity [8]. Hydrolysis and fermentation processes can be designed in various configurations, being performed separately, known as separate hydrolysis and fermentation (SHF) or simultaneously, known as simultaneous saccharification and fermentation (SSF) processes.

When hydrolysis of pretreated cellulosic biomass is performed with enzymes, these biocatalysts (endoglucanase, exoglucanase, and β-glucosidase) can be strongly inhibited by hydrolysis products, such as glucose, xylose, cellobiose, and other oligosaccharides. Therefore, SSF plays an important role to circumvent enzyme inhibition by accumulation of these sugars. Moreover, because accumulation of ethanol in the fermenters does not inhibit cellulases as much as high concentrations of sugars, SSF stands out as an important strategy for increasing the overall rate of cellulose to ethanol conversion. Some inhibitors present in the liquid fraction of the pretreated lignocellulosic biomass also have a significant and negative impact on enzymatic hydrolysis. Due to the decrease in sugar inhibition during enzymatic hydrolysis in SSF, the detoxifying effect of fermentation, and the positive effect of some inhibitors present in the pretreatment hydrolysate (e.g. acetic acid) on the fermentation, SSF can be an advantageous process when compared to SHF [8]. Another important advantage is a reduction in the sensitivity to infection in SSF when compared to SHF. However, this was demonstrated by Stenberg and co-authors [12] that this is not always the case. It was observed that SSF was more sensitive to infections than SHF. A major disadvantage of SSF is the difficulty in recycling and reusing the yeast since it will be mixed with the lignin residue and recalcitrant cellulose.

In SHF, hydrolysis of cellulosic biomass and the fermentation step are carried out in different units [8], and the solid fraction of pretreated lignocellulosic material undergoes hydrolysis (saccharification) in a separate tank by addition of acids/alkali or enzymes. Once hydrolysis is completed, the resulting cellulose hydrolysate is fermented and the sugars converted to ethanol. *S. cerevisiae* is the most employed microorganism for fermenting the hydrolysates of lignocellulosic biomass. This yeast ferments the hexoses contained in the hydrolysate but not the pentoses. Several strategies (screening biodiversity, metabolic and evolutionary engineering of microorganisms) have been attempted to overcome this metabolic limitation.

One of the main features of SHF process is that each step can be performed at its optimal operating conditions (especially temperature and pH) as opposed to SSF [7].Therefore, in SHF each step can be carried out under optimal conditions, i.e. enzymatic hydrolysis at 45-50°C and fermentation at 30-32°C. Additionally, it is possible to run fermentation in continuous mode with cell recycling. The major drawback of SHF, as mentioned before, is that the sugars released during hydrolysis might inhibit the enzymes. It must be stressed out that ethanol produced can also act as an inhibitor in hydrolysis but not as strongly as the sugars. A second advantage of SSF over SHF is the process integration obtained when hydrolysis and fermentation are performed in one single reactor, which reduces the number of reactors needed.

Experimental data from ethanol output using sugarcane bagasse as the substrate is being released in the literature. Vásquez et al. (2007) [13] described a process in which 30 g/L of ethanol was produced in 10h fermentation by *S. cerevisiae* (baker's yeast) at an initial cell concentration of 4 g/L (dry weight basis) using non-supplemented bagasse hydrolysate at 37°C. The hydrolysate was obtained by a combination of acid/alkali pretreatment, followed by enzymatic hydrolysis. Krishnan et al. (2010) [14] reported 34 and 36 g/L of ethanol on AFEX-treated bagasse and cane leaf residue, respectively, using recombinant *S. cerevisiae* and 6% (w/w) glucan loading during enzymatic hydrolysis. Overall the whole process produced around 20kg of ethanol per 100kg of each bagasse or cane leaf, and was performed during 250h including pretreatment, hydrolysis and fermentation. According to the authors this is the first complete mass balance on bagasse and cane leaf.

In view of the growing concern over climate change and energy supply, biofuels have received positive support from the public opinion. However, growing concern over first generation biofuels in terms of their impact on food prices and land usage has led to an increasing bad reputation towards biofuels lately. The struggle of 'land vs fuel' will be driven by the predicted 10 times increase in biofuels until 2050. The result is that biofuels are starting to generate resistance, particularly in poor countries, and from a number of activist non-governmental organizations with environmental agendas. This is highly unfortunate as it is clear that liquid biofuels hold the potential to provide a more sustainable source of energy for the transportation sector, if produced sensibly. Since replacement of fossil fuels will take place soon, a way to avoid these negative effects from first generation biofuels (mainly produced from potential food sources) is to make lignocellulosic derived fuels available within the shortest possible time. It is known that this process involves an unprecedented challenge, as the technology to produce these replacement fuels is still being developed [1]. Fuels derived from cellulosic biomass are essential in order to overcome our excessive dependence on petroleum for liquid fuels and also address the build-up of greenhouse gases that cause global climate change [2].

REFERENCES

1. L. D Gomez, C. G Steele-king, S. J Mcqueen-mason, Sustainable liquid biofuels from biomass: the writing's on the walls. New Phytologist 2008178473485
2. B Yang, C. E Wyman, Pretreatment: the key to unlocking low-cost cellulosic ethanol Biofuels Bioproducts and Biorefining 200822640
3. A Jagger, Brazil invests in second-generation biofuels. Biofuels, Bioproducts and Biorefining 20093810
4. M Galbe, G Zacchi, A review of the production of ethanol from softwood. Applied of Microbiology and Biotechnology 200259618628

5. A Pandey, C. R Socol, P Nigam, V. T Soccol, Biotechnological potential of agro-industrial residues. I: sugarcane bagasse 2000746980

6. J. R Mielenz, Ethanol production from biomass: technology and commercialization status. Current Opinion in Microbiology 20014324329

7. C. A Cardona, J. A Quintero, I. C Paz, Production of bioethanol from sugarcane bagasse: Status and perspectivesBioresource and Technology 201010147544766

8. S Kumar, S. P Singh, I. M Mishra, D. K Adhikari, Recent advances in production of bioethanol from lignocellulosic biomass Chemistry Engineering Technology 2009324517526Applied Biochemistry and Biotechnology 2001;91:5-21

9. S Takashima, A Nakamura, M Hidaka, H Masaki, T Uozumi, Molecular cloning and expression of the novel fungal β-glucosidase genes from Humicola grisea and Trichoderma reesei. Journal of Biochemistry (Tokyo) 1999125728736

10. Zhang YHPHimmel ME, Mielenz JR. Outlook for cellulase improvement: Screening and selection strategies Biotechnology Advances 2006452481

11. H Chen, S Jin, Effect of ethanol and yeast on cellulose activity and hydrolysis of crystalline cellulose Enzyme and Microbial Technology 20063914301432

12. K Stenberg, M Galbe, G Zacchi, The influence of lactic acid formation on the simultaneous saccharification and fermentation (SSF) of softwood to ethanol Enzyme Microbiology Technology 2000267179

13. P. M Vasquez, Silva JNC, Souza Jr MB, Pereira Jr N. Enzymatic hydrolysis optimization to ethanol production by simultaneous saccharification and fermentation. Applied Biochemistry and Biotechnology2007141 EOF53 EOF

14. C Krishnan, Sousa Lda C, Jin M, Chang L, Dale BE, Balan V. Alkali-based AFEX pretreatment for the conversion of sugarcane bagasse and cane leaf residues to ethanol Biotechnology and Bioengineering 201010744150

15. P. K Foreman, D Brown, L Dankmeyer, R Dean, S Diener, N. S Dunn-coleman, F Goedegebuur, T. D Houfek, G. J England, A. S Kelley, H. J Meerman, T Mitchell, C Mitchinson, H. A Olivares, Teunissen PJM, Yao J, Ward M. Transcriptional regulation of biomass-dedrading enzymes in the filamentous fungis Trichoderma reesei. The Journal of Biological Chemistry 2003278343198831997

CHAPTER 4

OBTAINING NEW CULTURES OF MICROORGANISMS THAT PRODUCES CELLULASES AND XYLANASES FROM THE SUGARCANE BAGASSE

LUDMYLLA NOLETO, DANIELLA MOREIRA AND FABRHCIA FARIA

4.1 BACKGROUND

The international energetic system strongly depends on fossil fuels, which causes negative effects in the environment, such as the global warming. Biofuels appear as an environmental and economic alternative for the energetic industry because of their potential source of renewable energy. Several studies are based on sugarcane culture and its derivatives, as bagasse, the sugarcane residue. Bioethanol can be produced by the fermentation of sugar or by the hydrolysis of cellulosic biomass [1]. The plant cell wall is constituted of cellulose (40-50%), hemicellulose (15-30%) and lignin (10-30%), forming the vegetal biomass. Cellulases are enzymes that form a complex that hydrolyses cellulosic materials, releasing sugars [2]. The main component of hemicellulose is the xylan, which is hydrolyzed by xylanases [3]. Cellulases as xylanases have a great biotechnological potential, they can be used in a variety of field: food, animal feed, textile and paper recycling industries. The sugarcane bagasse (SCB) is the most studied lignocellulosic waste for bioethanol production, because it is a by-product of conventional ethanol and can be find in large amount in Brazil [4]. Nowadays, the process of bioconversion of biomass has high cost and low specific activity of the enzymes that are necessary for the cellulose saccharification [5]. The aim of this research is to obtain microorganisms that hydrolyze the sugarcane bagasse and to quantify the sugar production.

4.2 METHODS

The microorganisms present in the SCB were isolated from three preparations: fresh SCB, SCB buried in soil for about 45 days and humid SCB—collected from two cane fields and stored in refrigerator. To obtain microorganisms, saline solution (NaCl 0,15 M) and rich medium (5 g/L peptone, 5 g/L NaCl and 10 g/L of SCB, pH 5.0 to 6.0) were used, followed by serial dilution. The selection medium contained cellulose and xylan and the enzymatic activity was visualized as a halo of hydrolysis around the culture, using congo red 1%. Submerged fermentation in minimum medium (MM) was used to induce cellulases and xylanases. The determination of enzymatic activity was measured by dinitrosalicilic acid (DNS), using the supernatants of culture as enzymes and Xylan birchwood-Sigma®, CMC-Sigma®, Avicel-Sigma® and Whatman paper filter as substrate for each enzymatic dosage.

4.3 RESULTS AND CONCLUSIONS

Seven cultures were selected (A3, B3, M2, M3, X7, F4 and D2) according to the halo of hydrolysis diameter to determine the enzymatic activity. The culture A3 proved to be a good producer of xylanase. The culture M3 produced cellulases with FPase and CMCase activity, showing that is good for cellulose hydrolysis. The culture X7 simultaneously produces cellulases and xylanases, which favors the hydrolysis of cellulose and hemicellulose using SCB as substrate. Although the activity of avicelase has no results, we cannot conclude that the enzyme was not produced; the microorganisms need to be induced by different specifications.

REFERENCES

1. Ogeda TL, Petri DFS: Hidrólise Enzimática de Biomassa. Quím Nova 2010, 33(7):1549-1558.
2. Bayer EA, Lamed R: The cellulose paradox: pollutant par excellence and/or a reclaimable natural resource? Biodegradation 1992, 3(2-3):171-188.
3. Sales MR, Moura RB, Macedo GR, Porto ALF: Variáveis que influenciam a produção de celulases e xilanase por espécies de Aspergillus. Pesquisa Agropecuária Brasileira 2010, 45(11):1290-1296.
4. Melo GR: Produção de celulas e xilanases pelo fungo termofílio Humicola grisea var. thermoidea em diferentes substratos lignocelulósicos. Dissertação (Mestrado), Instituto de Ciencias Biologicas, Universidade Federal de Goiás, Goiânia 2010.
5. Mayrink MICB: Produção de Enzimas Fúngicas e Avaliação do Potencial das Celulases na Sacarificação da Celulose. Dissertação (Mestrado), Universidade Federal de Viçosa, Viçosa, MG 2010.

CHAPTER 5

DESIGN AND OPTIMIZATION OF ETHANOL PRODUCTION FROM BAGASSE PITH HYDROLYSATE BY A THERMOTOLERANT YEAST *KLUYVEROMYCES* SP. IIPE453 USING RESPONSE SURFACE METHODOLOGY

DIPTARKA DASGUPTA, SUNIL KUMAR SUMAN,
DIWAKAR PANDEY, DEBASHISH GHOSH, RASHMI KHAN,
DEEPTI AGRAWAL, RAKESH KUMAR JAIN,
VASANTA THAKUR VADDE, AND DILIP K. ADHIKARI

5.1 INTRODUCTION

The global scenario demonstrates that lion share of research in past three decades have been focused on technological know-how development for bioethanol since its emergence as a potential fuel additive. All the key challenges on energy and economic front have already been pinpointed in various forums as sole restrictors for commercialization of lignocellulosic bioethanol technology (Cardona et al. 2010). Evidently a non-molasses

Design and Optimization of Ethanol Production from Bagasse Pith Hydrolysate by a Thermotolerant Yeast Kluyveromyces *sp. IIPE453 Using Response Surface Methodology.* © *Dasgupta et al.;* Springerplus *2013; 2:159. Licensee Springer. 2013 This article is published under license to BioMed Central Ltd. Creative Commons Attribution License (http://creativecommons.org/licenses/by/2.0).*

feedstock was to be brought into reality to meet excess ethanol demand for 5-10% compulsory blending. Biomass being a cheap and renewable raw material with abundant availability (Saxena et al. 2009; Kumar et al. 2009a; Cheng et al. 2008), has been considered as an excellent feedstock for bioethanol production due to its high holocellulosic content.

Ethanol production via fermentation route comprises of a series of biochemical reactions with numerous factors involved in the process. Conversion of lignocellulosic sugar hydrolysate into ethanol requires many other micro and macro elements apart from fermentable nitrogen which in right balance can always give optimum product yield. Statistical screening in this context provides a rapid assessment of key process variables in a systematic way whereby a perfect strategy can be materialized to improve targeted product yield. Response surface methodology (RSM) explores the relationships between several explanatory operating variables and one or more response variables and has been widely applied for optimization of ethanol production from various substrates (Uncu & Cekmecelioglu 2011; Jargalsaikhan & Saraçoğlu 2009).

In this paper, we have carried out RSM study of ethanol fermentation with thermotolerant yeast *Kluyveromyces* sp. IIPE453 (MTCC 5314) (Kumar et al. 2009b) to find optimum conditions for maximizing ethanol production via two step approach. Initial screening of factors were performed with Plackett-Burman Design (PBD) method to identify crucial parameters (Dong et al. 2012; Maruthai et al. 2012) affecting ethanol yield and to the degree based on their individual effect and interactions through Box-Behnken Design (BBD) technique (Mei et al. 2011; Palukurty et al. 2008). Further, an optimization study was conducted to maximize ethanol yield in shake flask. Optimized data has also been evaluated at bench scale bioreactor of 2 L working volume. One of the unique characteristics of *Kluyveromyces* sp. IIPE453 is its ability in utilizing pentose sugar for growth and fermentation with hexose sugar. Yeast cell biomass was grown with pentose rich fractions obtained after acid pretreatment of sugarcane bagasse (SCB) pith and fermented with glucose rich broth obtained after enzymatic saccharification of the pretreated pith.

An average sugarcane bagasse contains 35% pith and with 60% depithing efficiency, around 20% pith is removed during depithing operation either at sugar mill site or at paper mill premises. An average 300 tpd (Tonne per day) bagasse based paper mill generates 160 tpd pith (Jain et al. 2011). Pretreated pith has been utilized as substrate for production of single cell proteins (Rodriguez-Vazquez et al. 1992) as well for preparation of activated carbon for dye removal from aqueous solutions (Amin 2008). However, utilization vs. generation ratio is almost negligible. Even after using as a boiler fuel in paper industry itself (calorific value of 17.07 Kcal/Kg (Diez et al. 2011)) huge amount of pith remains unutilized and poses serious waste disposal problem. Bagasse based paper mills in India annually generates 45 – 55 million tons pith with a biochemical composition of holocellulose (68–69% w/w) including hemicellulose (20-21% w/w), lignin (21–22% w/w) and ash (6–7% w/w) (Sanjuan et al. 2001) which can be effectively used for ethanol production and thereby value addition to waste.

This study was conducted based on using this feedstock in order to integrate the process in a bio refinery mode attached with a sugar or paper and pulp industry and probably the first paper of this kind to the best of our knowledge.

5.2 MATERIALS AND METHODS

5.2.1 MATERIALS

Bagasse pith sample was generated in a depither unit with 100 mesh size at Central Pulp and Paper Research Institute (CPPRI), Saharanpur, Uttar Pradesh and used at CSIR-IIP, Dehradun for hydrolysis and saccharification. SCB pith was treated with steam and acid (8% w/w H_2SO_4) in a solid vs. liquid ratio of 1:10 at 120°C. for 90 minutes to extract pentose rich fraction (20 g/L) which was used as carbon source for cell biomass generation. Pretreated pith devoid of pentosans was further enzymatically saccharified using commercially available cellulase (Advanced Biochemicals Ltd, Mumbai, India) to get hexose rich stream (40 g/L) for ethanol fermentation. Saccharification was carried out using 7% w/w of enzyme with solid vs. liquid ratio of 1:10 at 50°C. for 22 h.

5.2.2 MICRO-ORGANISM AND CULTURE CONDITIONS

Kluyveromyces sp. IIPE453 (MTCC 5314), a thermophilic yeast (optimum growth temperature 45°C.) isolated from dumping sites of crushed SCB in a local sugar mill was used in this experimental study. The stock culture was maintained on YPD agar medium (composition in g/L; yeast extract, 10.0; peptone, 20.0; dextrose 20.0; agar agar 20.0; pH 4.5-5.0).

5.2.3 EXPERIMENTAL DESIGN

Growth was carried out in prehydrolysate (pentose broth) at 45°C. for 16 h. Nutrient screening and optimization for ethanol production were performed at same temperature in shake flasks (80 ml working volume) in the hydrolysate (hexose broth) supplemented with various nutrients according to experimental design. The physical parameters, pH and fermentation time were maintained as per design specifications (Table 1). All experiments were carried out in replicates and results are reported in terms of mean values. Experimental design and statistical analyses were done using *Reliasoft* Design of Experiment (DOE) software with risk factor (α) values of 0.05 (95% level of confidence) for PBD and 0.01 (99% level of confidence) for BBD. Criterion of predicted model acceptance was based on their adjusted coefficient of regression (Radj 2) with value above 0.95. Variables with P values lower than 0.05 (PBD) and 0.01 (BBD) were considered to have significant effect on the response.

5.2.3.1 PLACKETT BURMAN DESIGN

A two level PBD (Plackett & Burman 1946) experimental matrix was set up to identify the factors and estimate their significance in ethanol production. It predicts linear model where only main effects are taken into consideration.

$$Response = a + \sum b_i * X_i \tag{1}$$

TABLE 1: Plackett burman design for screening of factors.

Serial	Run order	pH	Inoculum volume (%v/v)	Substrate conc. (g/L)	Yeast extract (g/L)	$MgSO_4$ (g/L)	$(NH_4)_2SO_4$ (g/L)	Na_2HPO_4 (g/L)	KH_2PO_4 (g/L)	Fermentation time (h)	Response variable ethanol conc. (g/L) Experimental	Response variable ethanol conc. (g/L) Model predicted
1	1	+1	+1	-1	+1	+1	+1	-1	-1	-1	7.20	6.91
2	6	-1	-1	-1	+1	-1	+1	+1	-1	+1	8.05	8.14
3	3	+1	-1	+1	+1	-1	+1	+1	+1	-1	7.18	7.18
4	2	-1	+1	+1	-1	+1	+1	+1	-1	-1	8.07	8.35
5	4	-1	+1	-1	-1	+1	-1	+1	+1	+1	7.86	7.76
6	10	-1	+1	+1	+1	-1	-1	-1	+1	-1	9.03	9.13
7	11	+1	-1	+1	+1	+1	-1	-1	-1	+1	10.09	10.37
8	8	+1	+1	-1	-1	+1	+1	-1	+1	+1	8.05	8.33
9	9	+1	+1	+1	-1	-1	-1	+1	-1	+1	7.30	7.02
10	12	-1	-1	-1	-1	-1	-1	-1	-1	-1	8.20	8.10
11	5	-1	-1	+1	-1	+1	+1	-1	+1	+1	13.40	13.12
12	7	+1	-1	-1	-1	+1	-1	+1	+1	-1	5.60	5.69

TABLE 2: Factors with their coded levels.

Serial number	Variable	Low (-1)	Center point (0)	High (+1)
1	pH	4.5	5	5.5
2	Fermentation time (h)	24	36	48
3	Substrate Concentration (g/L)	20	30	40
4	Yeast extract (g/L)	1	2.5	5
5	$MgSO_4$ (g/L)	0.06	0.09	0.12
6	$(NH_4)_2SO_4$ (g/L)	1	3	5
7	Na_2HPO_4 (g/L)	0.15	0.30	0.45
8	KH_2PO_4 (g/L)	0.15	0.30	0.45
9	Inoculum volume (% v/v)	5	7.5	10

Response indicates dependent variable in terms of overall ethanol production (g/L), a being the model intercept. X_i represents different levels of independent variables with b_i coefficients as predicted by the equation. In this paper, 9 independent variables were selected, e.g. physical parameters such as pH, fermentation time, inoculum volume (%v/v) and media components such as sugar concentration, yeast extract, magnesium sulphate [$MgSO_4$], ammonium sulphate [$(NH_4)_2SO_4$], di-sodium hydrogen phosphate [Na_2HPO_4] and potassium di-hydrogen phosphate [KH_2PO_4]. Table 1 illustrates the design matrix of various components with coded values; low (-1) and high (+1) while Table 2 represents their actual values. Pareto charts were plotted to highlight most significant factors responsible for ethanol production.

5.2.3.2 BBD DESIGN AND OPTIMIZATION

BBD technique is a statistical indication of quadratic effect factors obtained after initial factorial screening studies and their interactions (Box & Behnken 1960). Based on Pareto chart results, BBD matrix was constructed with four significant factors (substrate concentration, pH, fermentation time and Na_2HPO_4 concentration) each having 3 levels (-1, 0 and 1) with 27 experimental designs as shown in Table 3. Rest non-significant factors namely Inoculum volume, Yeast extract, $MgSO_4$, $(NH_4)_2SO_4$ and KH_2PO_4 were maintained at their respective low level values (Table 2). A second order polynomial model was predicted with DOE (equation 2) indicating linear, interaction and quadratic effect of variables on system response as either + ve or -ve. ANOVA analysis of the model was performed to evaluate its statistical significance.

$$\begin{aligned}
Response = {} & a_1 + b_1 * A + b_2 * B + b_3 * C + b_4 * D \\
& + b_1 * A * B + b_6 * A * C + b_7 * A * D \\
& + b_8 * B * C + b_9 * B * D + b_{10} * C * D \\
& + b_{11} * A^2 + b_{12} * B^2 + b_{13} * C^2 + b_{14} * D^2
\end{aligned} \quad (2)$$

where, A, B, C and D are the independent variables, a_1 is an offset term, b_1 to b_4 are linear term coefficients, b_5 to b_{10} indicate interaction terms and b_{11} to b_{14} represent quadratic effect.

5.2.4 MODEL VALIDATION IN SHAKE FLASK AND SCALE UP STUDY IN BENCH SCALE BIOREACTOR

BBD predicted response led to identification of optimization conditions in terms of key independent variables having significant effect on system response. To validate authenticity of software generated model, fermentation was carried out in shake flask under optimized conditions and further tested on a 2 L bench scale bioreactor (NBS Bioflo 110) equipped with *supervisory control and data acquisition* (SCADA) system. Yeast biomass generated on prehydrolysate (20 g/L pentose conc.) was inoculated in fermentation broth having SCB pith hydrolysate (40 g/L glucose conc.) in shake flask as per model predicted optimized conditions as well as in 2 L NBS bioflo110.

TABLE 3: Box behnken design.

Run order	Random	Substrate conc. (g/L) (A)	pH (B)	Fermentation time (h) (C)	Na$_2$HPO$_4$ (g/L) (D)	Experimental ethanol conc. (g/L)	Model predicted ethanol conc. (g/L)
1	15	0	-1	0	+1	10.90	10.32
2	14	0	+1	0	-1	8.60	7.79
3	9	-1	0	-1	0	4.20	3.95
4	12	+1	0	+1	0	11.55	12.07
5	4	+1	+1	0	0	7.10	6.91
6	17	-1	0	0	-1	9.30	9.34
7	19	-1	0	0	+1	4.40	4.35
8	20	+1	0	0	+1	11.50	11.47
9	24	0	+1	+1	0	7.90	7.45
10	7	0	0	-1	+1	7.25	7.37
11	25	0	0	0	0	7.60	8.01
12	11	0	0	+1	0	8.32	7.65
13	2	+1	-1	0	0	14.60	13.53
14	13	0	-1	0	-1	12.70	12.61
15	8	0	0	+1	+1	8.00	8.45
16	10	+1	0	-1	0	8.10	8.37
17	23	0	-1	+1	0	12.00	12.27
18	3	-1	+1	0	0	3.60	4.29
19	1	-1	-1	0	0	7.50	7.31
20	21	0	-1	-1	0	7.90	8.57
21	5	0	0	-1	-1	6.80	7.04
22	6	0	0	+1	-1	12.80	13.36
23	18	+1	0	0	-1	11.00	11.06
24	26	0	0	0	0	8.00	8.01
25	16	0	+1	0	+1	5.40	5.50
26	27	0	0	0	0	7.70	8.01
27	22	0	+1	-1	0	4.10	3.75

5.2.5 ANALYTICAL METHODS

Sugar and ethanol concentration (g/L) was quantified by HPLC (UFLC Shimadzu) with PL Hiplex-H acid 8 µm column (100 × 7.7 mm diameter, by PL Polymer laboratory, UK). The column was eluted with a mobile phase 1 mM sulfuric acid at a flow rate of 0.7 ml/min at column oven temperature 57°C. with RI detector.

5.3 RESULTS AND DISCUSSION

5.3.1 EVALUATION OF KEY VARIABLES AFFECTING ETHANOL PRODUCTION

Lignocellulosic ethanol production requires various micro and macro elements along with fermentable sugar and nitrogen which in best commingle results in optimum product yield where a controlled environment is again a prerequisite (Asli 2009; Anupama et al. 2010). Magnesium, being the cofactor for glycolytic enzymes involved in fermentation (Lodolo et al. 2008) and potassium being the regulator of pH via K^+/H^+ transport system, (Kudo et al. 1998) are essential cations governing ethanol fermentation. Ammonium salts stimulate glucose fermentation by lowering induction period (Muntz 1947) and maintaining an optimum carbon to nitrogen (C/N) ratio. Substrate concentration primarily affects uptake rates and thereby product rate kinetics. High substrate concentration negatively hampers ethanol productivity leading to a lower titer due to repression of glycolytic enzymes (Bisson & Fraenkel 1984). Yeast extract is a rich source of vitamins and promotes cell growth and proliferation. Hence, the above mentioned variables have been chosen to screen and develop a low cost fermentation medium with optimum blend of nutrients and physical parameters for bioethanol production.

PBD identified the key variables among selected ones via *Pareto* chart illustrated in Figure 1. Factors such as pH, fermentation time, substrate concentration and Na_2HPO_4 with T values above threshold (4.30 in this case) and P values lower than 0.05 as represented by regression analysis (Table 4) had a substantial effect on ethanol yield and were considered for further evaluation by BBD, while rest of the variables did not have a meaningful contribution to ethanol production. Fermentation process is directly affected by the amount of viable cells present in broth. An optimum inoculum volume of 5% (v/v) was sufficient to carry out the fermentation process. Higher concentration of the same had no effect on ethanol yield improvement and thus was considered to be non-significant variable in the process (Figure 1). The model considering main effects (equation not shown) was found to be fairly accurate having a R^2 value of 0.98 with a R_{adj}^2 value of 0.92 with experimental and model predicted response being fairly close to each other.

5.3.2 OPTIMIZATION OF PHYSICAL PARAMETERS AND MEDIA COMPONENTS FOR ETHANOL PRODUCTION

BBD matrix with response is shown in Table 3. A second order polynomial model fit to the experimental data for optimizing ethanol production via response surface method (RSM) predicts response as a function of four variables and their interactions in terms of their coded values.

$$
\begin{aligned}
Response = 7.67 + 2.21 * A - 2.40 * B + 1.81 * C \\
-1.14 * D - 0.90 * A * B - 0.1675 * A * C \\
+1.35 * A * D - 0.07 * B * C \\
-0.35 * A * D - 0.07 * B * C \\
+0.34 * B^2 - 0.08 * C^2 + 1.13 * D^2
\end{aligned}
\tag{3}
$$

ANOVA calculations listed in Table 5 show that the model F and P values are 49.252 and 2.11×10^{-8}. This signifies the model with 99% level of confidence ($\alpha = 0.01$) and all effects namely linear, interaction and quadratic are exhibited. Quality of fit model was estimated by R_{adj}^2 and predicted R^2 (R_{pred}^2) values were found to be 0.96 and 0.90 respectively which are fairly high and accurate measures of precision (Ohtani 2000). This indicates that only 4% variation in response cannot be suitably explained by the model. Response values for each run calculated by developed model showed little or no variation compared to test results. This indicated that model equation very well corresponded to BBD experimental data. Statistical significance of the model term coefficients was determined by student's t-test and p test values as illustrated in Table 6. It was observed that main effects were significant for each of four coded factors whereas interactions among pH and substrate concentration, substrate concentration & Na_2HPO_4 concentration, fermentation time & Na_2HPO_4 concentration were important as indicated by their high T and low P values.

TABLE 4: Regression analysis for Plackett Burman design variables.

Term	Effect	Coefficient	Standard error	T value	P value
Intercept		8.3376	0.1481	56.296	0.0003
pH	-1.5322	-0.7661	0.1481	-5.1726	0.0354
% Inoculum (v/v)	-0.8345	-0.4173	0.1481	-2.8174	0.1063
Substrate conc.	1.6883	0.8441	0.1481	5.6997	0.0294
Yeast extract	-0.1995	-0.0997	0.1481	-0.6734	0.5701
Magnesium sulphate	0.7329	0.3665	0.1481	2.4744	0.1318
Ammonium sulphate	0.6458	0.3229	0.1481	2.1803	0.161
Sodium di-hydrogen phosphate	-1.984	-0.992	0.1481	-6.6982	0.0216
Potassium di-hydrogen phosphate	0.3705	0.1853	0.1481	1.251	0.3374
Fermentation time	1.5751	0.7875	0.1481	5.3174	0.0336

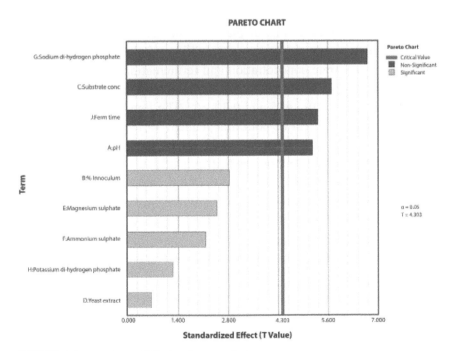

FIGURE 1: Pareto chart of Placket Burman design.

TABLE 5: ANOVA table for BBD model.

Source of variation	Degrees of freedom	Sum of squares [Partial]	Mean squares [Partial]	F ratio	P value
Model	14	211.4135	15.101	49.2852	2.11E-08
Linear Effects	4	185.1535	46.2884	151.072	3.75E-10
Interaction Effects	6	18.0454	3.0076	9.8158	0.0005
Quadratic Effects	4	8.2146	2.0537	6.7026	0.0045
Residual	12	3.6768	0.3064		
Lack of Fit	10	3.5901	0.359	8.2849	0.1124
Pure Error	2	0.0867	0.0433		
Total	26	215.0903			

3D response surface graphs display the characteristic effects of key process variables on ethanol concentration. Figure 2 demonstrates the response against substrate concentration and fermentation time while other two factors namely pH and Na_2HPO_4 concentration are maintained at their centre point values (0,0), i.e. 5 and 0.30 g/L. The linear surface exhibits a greater first degree effect of both independent variables on system response. An increase in both factors lead to enhanced ethanol yield, maximum being 11.77 g/L at 48 hour and 40 g/L of substrate concentration and decrease in ethanol yield on reduction of the same. Thus, both factors have a positive effect on the dependent variable. On the contrary, effects of Na_2HPO_4 concentration and pH at hold values for substrate concentration (30 g/L) and fermentation time (36 h) illustrate that system should be maintained at low values for both variables to attain maximum ethanol production (Figure 3).

The surface is more concave nature in this case compared to Figure 2 and represents quadratic and interaction effects in addition to linear ones with maximum ethanol concentration of 12.44 g/L at pH 4.5 and Na_2HPO_4 concentration 0.15 g/L. Effect of substrate concentration and

TABLE 6: Significance of term coefficients for BBD.

Term	Coefficient	Standard error	T value	P value
Intercept	7.7667	0.3196	24.3025	1.42E-11
A	2.2108	0.1598	13.8358	9.75E-09
B	-2.4083	0.1598	-15.0717	3.68E-09
C	1.8517	0.1598	11.588	7.13E-08
D	-1.1458	0.1598	-7.1708	1.13E-05
A*B	-0.9	0.2768	-3.2518	0.0069
A*C	-0.1675	0.2768	-0.6052	0.5563
A*D	1.35	0.2768	4.8778	0.0004
B*C	-0.075	0.2768	-0.271	0.791
B*D	-0.35	0.2768	-1.2646	0.23
C*D	-1.3125	0.2768	-4.7423	0.0005
A²	0.1996	0.2397	0.8327	0.4213
B²	0.3408	0.2397	1.422	0.1805
C²	-0.0817	0.2397	-0.3407	0.7392
D²	1.1346	0.2397	4.7336	0.0005

A: Substrate concentration (g/L)
B: pH
C: Fermentation time (h)
D: Na_2HPO_4 Conc. (g/L)

Na_2HPO_4 concentration on ethanol production at a fixed pH value of 5 and 36 h fermentation time is depicted in Figure 4. It demonstrates that Na_2HPO_4 and substrate concentration at their maximum values 0.45 g/L and 40 g/L respectively, lead to maximum ethanol production of 11.51 g/L whereas Na_2HPO_4 concentration at its lowest value (0.15 g/L) with same substrate concentration yields almost same ethanol (11.15 g/L). Hence, ethanol production is more sensitive to changes in substrate concentration compared to Na_2HPO_4 concentration when other two variables pH and fermentation time are fixed at their midpoint values. However, interaction effects between these two are + ve and statistically significant as predicted by the model equation for considerable ethanol yield. Based on polynomial model, optimization study was carried out for maximizing ethanol

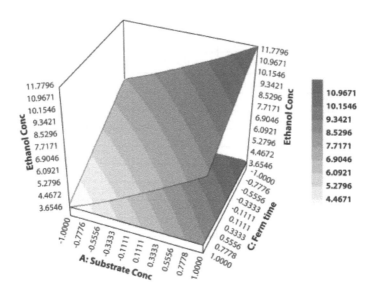

FIGURE 2: Effects of substrate concentration and fermentation time on ethanol production hold values: B = 0 (pH = 5.0), D = 0 (Na$_2$HPO$_4$ = 0.15 g/L).

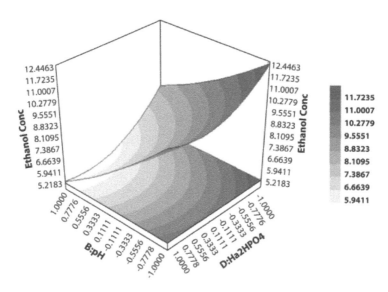

FIGURE 3: Effects of pH and Na$_2$HPO$_4$ concentration on ethanol production hold values: A =0 (substrate Conc. = 30 g/L), C =0 (fermentation time = 36 h).

production. Maximum ethanol concentration predicted by the model was found to be 17.39 g/L with 40 g/L (+1) substrate concentration, pH 4.5 (-1), 48 h (+1) fermentation time and 0.15 g/L (-1) Na_2HPO_4 (Figure 5). The data was further validated in a shake flask study where the experiment was carried out under optimized condition.

5.3.3 SCALE UP STUDY IN BENCH SCALE BIOREACTOR

Scale up study was conducted in bioreactor with optimized conditions yielding 17.44 g/L of ethanol which is almost identical to the model predicted value with residual hexose concentration of 1.2 g/L in the hydrolysate. This validated the accuracy of predicted model and confirmation of an optimum point within system for achieving targeted ethanol production. The ethanol yield (Yp/S) in terms of consumed sugar was 88% of theoretical value with specific ethanol productivity of 0.36 g/L/h.

5.4 CONCLUSION

Identification and optimization of key process variables for ethanol production from SCB pith hydrolysate could successfully be achieved using PBD and RSM. Four variables namely Substrate Conc., pH, fermentation time and Na_2HPO_4 were most significant factors affecting ethanol production. Final ethanol concentration and yield attained under optimum fermentation conditions was 17.44 g/L and 88% of theoretical value which was identical to the model predicted response. The ethanol yield, productivity and fermentation conditions for ethanol production from SCB pith via this process was compared with other other lignocellulosic bioethanol processes (Table 7) with different fermenting strains. The ethanol yield obtained with the current process is found to be significantly high in comparison to other processes utilizing different lignocellulosic/waste feedstocks for bioethanol production.

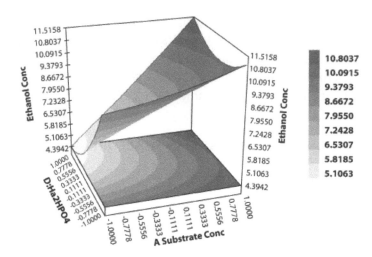

FIGURE 4: Effects of substrate concentration and Na_2HPO_4 concentration on ethanol production hold values: B =0 (pH = 5.0), C =0 (fermentation time = 36 hours).

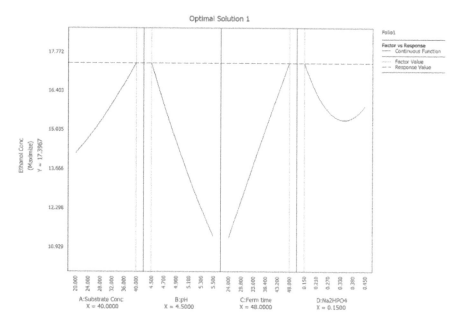

FIGURE 5: Optimization conditions for maximizing ethanol yield predicted by reliasoft DOE.

TABLE 7: Comparative analysis of different ethanol processes with lignocellulosic/waste material.

SI #	Lignocellulosic/waste raw material	Strain used	Temperature (°C)	pH	Sugar conc. (g/L)	Ethanol conc.(g/L)	Inoculum volume (%v/v)	Yield (% theoretical)	Productivity (g/L/h)	Reference
1	Bagasse pith hydrolysate	*Kluyveromyces* sp. IIPE453	45	4.5	40	17.40	5	88	0.36	this paper
2	Softwood	*Pichiastipitis* CBS6054: *S. cerevisiae* Y5	30	5.0	65	27.40	-	85.1	0.28	(Wan et al. 2012)
3	Kinnow waste and banana peels	*Consortia of Pachysolen-tannophilus* (MTCC 1077) and *S. cerevisiae* G	30	-	63	26.84	4:6	83.5	0.55	(Han et al. 2011)
4	Sugarcane bagasse	*Kluyveromyces fragilis*	35	5.5	180	32.60	-	36	0.45	(Sharma et al. 2007)
5	Korean food waste	*S. cerevisiae* 7904	35	5.4	75	24.17	2.5	63	0.60	(Sasikumar & Virutha-giri 2008)
6	Tapioca stem	*Fusarium oxysporum*	30	5.5	33	8.64	2	51.33	0.05	(Man et al. 2010)
7	Miscanthus biomass	*S. cerevisiae*	32	-	140	59.20	7	83.92	1.23	(Magesh et al. 2011)

REFERENCES

1. Amin NK: Removal of reactive dye from aqueous solutions by adsorption onto activated carbons prepared from sugarcane bagasse pith. Desalination 2008, 223:152–161.
2. Anupama MP, Mahesh DG, Ayyanna C: Optimization of fermentation medium for the production of ethanol from jaggery using box-behnken design. Int J Appl Biol Pharm Technol 2010, 1:34–45.
3. Asli MS: A study on some efficient parameters in batch fermentation of ethanol using Saccharomyces cerevesiae SC1 extracted from fermented siahesardasht pomace. Afr J Biotechnol 2009, 9:2906–2912.
4. Bisson LF, Fraenkel DG: Expression of kinase dependent glucose uptake in Saccharomyces cerevisae . J Bacteriol 1984, 159:1013–1017.
5. Box GEP, Behnken DW: Some new three level designs for the study of quantitative variables. Technometrics 1960, 2:455–475.
6. Cardona CA, Quintero JA, Paz IC: Production of bioethanol from sugarcane bagasse: status and perspectives. Bioresour Technol 2010, 101:4754–4766.
7. Cheng KK, Cai BY, Zhang JA, Ling HZ, Zhou YJ, Ge JP, Xu JM: Sugarcane bagasse hemicellulose hydrolysate for ethanol production by acid recovery process. Biochem Eng J 2008, 38:105–109.
8. Diez OA, Càrdenas GJ, Mentz LF: Calorific value of sugarcane bagasse, pith and their blends (Tucumàn, Argentine Republic). Rev indagric Tucumán 2011, 87:29–38.
9. Dong Y, Zhang N, Lu J, Lin F, Teng L: Improvement and optimization of the media of Saccharomyces cerevisae strain for high tolerance and high yield of ethanol. Afr J Biotechnol 2012, 6:2357–2366.
10. Han M, Choi G-W, Kim Y, Koo B-C: Bioethanol production by miscanthus as a lignocellulosic biomass: focus on high efficiency conversion to glucose and ethanol. Bioresources 2011, 6:1939–1953.
11. Jain RK, Thakur VV, Pandey D, Adhikari DK, Dixit AK, Mathur RM: Bioethanol from bagasse pith a lignocellulosic waste biomass from paper/sugar industry. IPPTA 2011, 23:169–173.
12. Jargalsaikhan O, Saraçoğlu N: Application of experimental design method for ethanol production by fermentation of sunflower seed hull hydrolysate using Pichiastipitis NRRL-124. Chem Eng Comm 2009, 196:93–103.
13. Kudo M, Vagnoli P, Bisson LF: Imbalance of pH and potassium concentration as a cause of stuck fermentations. Am J Enol Vitic 1998, 49:295–301.
14. Kumar P, Barett DM, Delwiche MJ, Stroeve P: Methods for pretreatment of lignocellulosic biomass for efficient hydrolysis and biofuel production. Ind Eng Chem Res 2009, 48:3713–3729.
15. Kumar S, Singh SP, Mishra IM, Adhikari DK: Ethanol and Xylitol production from glucose and xylose at high temperature by Kluyveromyces sp. IIPE453. J Ind Microbiol Biotechnol 2009, 36:1483–1489.
16. Lodolo EJ, Kock JLF, Axcell BC, Brooks M: The yeast Saccharomyces cerevisiae - the main character in beer brewing. Fems Yeast Res 2008, 8:1018–1036.

17. Magesh A, Preetha B, Viruthagiri T: Statistical optimization of process variables for direct fermentation of 226 white rose tapioca stem to ethanol by fusarium oxysporum. Int J Chem Tech Res 2011, 3:837–845.

18. Man HL, Behera SK, Park HS: Optimization of operational parameters for ethanol production from korean food waste leachate. Int J EnvSci Tech 2010, 7:157–164.

19. Maruthai K, Thangavelu V, Kanagasabai M: Statistical screening of medium components on ethanol production from cashew apple juice using Saccharomyces Diasticus . Int J Chem Biol Engg 2012, 6:108–111.

20. Mei X, Liu R, Shen F, Cao W, Wu H, Liu S: Optimization of solid-state ethanol fermentation with the soluble carbohydrate in sweet sorghum stalks using response surface methodology. J Biobased Mater Bio 2011, 5:532–538.

21. Muntz JA: The role of potassium and ammonium ions in alcoholic fermentation. J Biol Chem 1947, 17:653–665.

22. Ohtani K: Bootstrapping R 2 and adjusted R 2 in regression analysis. Econ Model 2000, 17:473–483.

23. Palukurty MA, Telgana NK, Bora HSR, Mulampaka SN: Screening and optimization of metal ions to enhance ethanol production using statistical experimental designs. Afr J Microbio Res 2008, 2:87–94.

24. Plackett RL, Burman JP: The design of optimum multifactorial experiments. Biometrika 1946, 33:2147–2157.

25. Rodriguez-Vazquez R, Villanueva-Ventura G, Rios-Leal E: Sugarcane bagasse pith dry pretreatment for single cell protein production. Bioresour Technol 1992, 39:17–22.

26. Sanjuan R, Anzaldo J, Vargas J, Turrado J, Patt R: Morphological and chemical composition of pith and fibres from mexican sugarcane bagasse. HolzalsRoh-Und Workestoff 2001, 59:447–450.

27. Sasikumar E, Viruthagiri T: Optimization of process conditions using response surface methodology (rsm) for ethanol production from pretreated sugarcane bagasse: kinetics and modeling. Bioenerg Res 2008, 1:239–247.

28. Saxena RC, Adhikari DK, Goyal HB: Biomass-based energy fuel through biochemical routes: a review. Renewab Susta Ener Rev 2009, 13:167–169.

29. Sharma N, Kalra KL, Oberoi HS, Bansal S: Optimization of fermentation parameters for production of ethanol from kinnow waste and banana peels by simultaneous saccharification and fermentation. Indian J Microbiol 2007, 47:310–316.

30. Uncu ON, Cekmecelioglu D: Cost-effective approach to ethanol production and optimization by response surface methodology. Waste Manage 2011, 31:636–643.

31. Wan P, Zhai D, Wang Z, Yang X, Tain S: Ethanol production from nondetoxified dilute-acid lignocellulosichydrolysate by cocultures of Saccharomyces cerevisiae Y5 and Pichiastipitis CBS6054. Biotechnol Res Int 2012, 1–6.

CHAPTER 6

ULTRA-STRUCTURAL MAPPING OF SUGARCANE BAGASSE AFTER OXALIC ACID FIBER EXPANSION (OAFEX) AND ETHANOL PRODUCTION BY *CANDIDA SHEHATAE* AND *SACCHAROMYCES CEREVISIAE*

ANUJ K. CHANDEL, FELIPE F. A. ANTUNES, VIRGILIO ANJOS, MARIA J. V. BELL, LEONARDE N. RODRIGUES, OM V. SINGH, CARLOS A. ROSA, FERNANDO C. PAGNOCCA, AND SILVIO S. DA SILVA

6.1 BACKGROUND

Multidimensional applications have increased the demand for fossil fuels, sending disturbing signals regarding low levels of crude oil underground and the burden on aging refineries [1]. Recent developments have explored marginally meaningful alternatives to fossil fuels such as bioethanol using corn grains or sugarcane juice as substrates, which would result in tremendous price hikes for basic food commodities around the world [2]. However, bioethanol produced from sustainable feedstocks such as lignocellulosics has drawn worldwide attention as a legitimate alternative to gasoline. The development of green gasoline or ethanol from abundantly available cellulosic materials in nature is gaining significant momentum as a sustainable mitigation strategy [3]. As an alternative to fossil and food-based fuels, cellulosic ethanol offers near-term environmental sustainability benefits. However, the selection of a raw substrate and its efficient utilization are critical steps in the process of economization. Brazil and India are the largest producers of sugarcane, with an annual production of about 650 and 350 MMT/year respectively [4], able to supply sugarcane bagasse (SB) in abundance round the year.

SB consists of crystalline cellulose nanofibrils embedded in an amorphous matrix of cross-linked lignin and hemicelluloses that impairs enzyme and microbial accessibility [5]. Structural changes in the cellular components of SB have been studied after pretreatment with formic acid [6], sono-assisted acid hydrolysis [7], and sequential acid–base [8] methodologies. However, visual characterizations of cellular components elucidating the hemicellulose degradation and delocalization of lignin after oxalic acid fiber expansion (OAFEX) pretreatment have yet to be performed. Cell wall anatomy of SB and macroscopic/microscopic barriers for cellulase-mediated saccharification reveal that cell wall hydrolysis encompasses several orders of magnitude (100^{-10-9} meters [9,10].

Oxalic acid (OA) is a strong dicarboxylic acid with higher catalytic efficiency than the sulfuric acid, and acts primarily upon hemicellulose. Pretreatment with OA leaves cellulignin in a fragile form, making it amenable to concerted cellulolytic enzyme action on cellulose, which yields glucose. The current pretreatment methodologies used to degrade the holocellulosic (cellulose + hemicellulose) fraction of the plant cell wall have economic and environmental limitations. However, OAFEX has been found to be an efficient pretreatment strategy for hemicellulose removal from giant reed [11], *Saccharum spontaneum* [12], corn-cobs [13], and wood chips [14]. Therefore, a comprehensive structural analysis of the nanoscale architecture of OAFEX-treated cell walls in tandem with molecular changes will assist to explore the fundamental mechanisms of biomass recalcitrance.

The fermentation of pentose and hexose sugars present in lignocellulose hydrolysate is important to produce "economic ethanol" [15,16]. *Candida shehatae*, native xylose-fermenting yeast, has shown the capability to utilize pentose sugars efficiently for ethanol production [17,18].

The microorganism *Saccharomyces cerevisiae* is a perennial choice for bio-ethanol production from glucose-rich cellulosic hydrolysates [19]. We attempted to pretreat SB using OA followed by enzymatic hydrolysis. The released sugars (12.56 g/l xylose and 1.85 g/l glucose) were than subjected to ethanol fermentation using *C. shehatae* UFMG HM 52.2 and *S. cerevisiae* 174 in batch fermentation. A multiscale structural analysis of native, OAFEX-treated and enzyme-digested SB was performed using scanning electron microscopy (SEM), atomic force microscopy (AFM), X-ray diffraction (XRD), Raman spectroscopy, near infrared spectroscopy (NIR), and Fourier transform infrared (FTIR) analysis, and revealed the structural differences before and after OAFEX treatment and enzymatic hydrolysis.

6.2 RESULTS AND DISCUSSION

The abundance availability of SB as substrate can be used for fuel ethanol production without jeopardizing food and feed production. SB is rich in lignocellulosics contained (% d. wt.): cellulose 45.0, hemicellulose 25.8, lignin 19.1, structural ash 1.0, extractive 9.1 [20]. The holocellulosic (hemicellulose + cellulose) content of SB (circa 70%) is fairly comparable with other lignocellulosic materials studied for ethanol production, such as wheat straw (54%), birch (73%), spruce (63.2%) (11), corn stover (59.9%), and poplar (58.2%) [21]. Biomass recalcitrance and efficient sugar conversion into ethanol are among the key hindrances to ethanol production that are preventing biorefineries from taking a central role in the energy sector.

6.2.1 OAFEX TREATMENT OF SUGARCANE BAGASSE AND DETOXIFICATION OF HEMICELLULOSIC HYDROLYSATE

OA is a strong organic acid, and due to its di-carboxylic properties, is better for hemicellulose hydrolysis than mineral acids such as sulfuric or hydrochloric acid [13]. OAFEX pretreatment is known for its precise action to degrade hemicellulose with fewer inhibitors and increased surface area of cellulose for improved enzymatic action [11,13,14,22,23].

The dilute OA hydrolysis of SB (160°C; 3.5% OA w/v; 10% total solids; 20 min residence time) depolymerized the hemicellulose fraction of the bagasse cell wall into sugars (xylose 12.56 g/l, glucose 1.85 g/l, arabinose 0.85 g/l) with the hemicellulose conversion (90.45%) and inhibitors (0.234 g/l furfurals, 0.103 g/l 5-hydroxy methyl furfural (5-HMF), 1.47 g/l acetic acid and 2.95 g/l total phenolics) (Table 1). Diluted OA hydrolysis degrades the hemicellulosic fraction of the plant cell wall into its monomeric constituents, such as xylose and other sugars, in addition to inhibitory compounds [13,22].

After hemicellulose is removed from the substrate during the OAFEX treatment, the leftover solid material, called cellulignin, becomes accessible for enzymatic saccharification. Critical factors such as lignocellulosic substrate, temperature, acid load, residence time, and substrate-to-liquid ratio play key roles in breaking down hemicellulose into its monomeric constituents [13,14], in addition to releasing inhibitors from the substrate [17]. Scordia et al. [12] reported 30.70 g/L xylose (93.73% conversion), 2.60 g/L glucose (54.17% conversion), 1.40 g/L arabinose, 3.60 g/L acetic acid, 0.68 g/L furfurals, 0.10 g/L HMF, and 6.58 g/L phenolics from *S. spontaneum* under the hydrolytic conditions (158°C, 16 min, 3.21% w/w OA, solid-to-liquid ratio of 1:4). The OA-mediated hydrolysis of giant reed (*Arundo donax L.*) revealed 100% recovery of xylose, arabinose, glucose, 6.55 g/L furfural, and 0.18 g/L HMF at defined hydrolysis conditions, i.e. 190°C, 25 min, 5% w/w OA [11]. Maize residues pretreated with OA (160°C, 1.8% OA, 10 min residential time) showed 28±2.5% w/w xylose yield [23]. These studies reveal the potential of OA pretreatment for hemicellulose degradation into fermentable sugars.

Fermentative inhibitors are common in acidic hydrolysis of SB. OA hydrolysate can be detoxified effectively using calcium hydroxide overliming: raising the pH of hydrolysate to 10.0. Regardless the source of availability of lignocellulosic hydrolysate, the detoxification using calcium hydroxide overliming is tedious and a subject of further investigation. It is an intensive step which cause precipitation and stirring problems. The re-adjustment to pH 6.0 showed efficient removal of inhibitors from the hydrolysate (Table 1). After $Ca(OH)_2$ overliming of the SB hemicellulosic hydrolysate, a significant reduction in furfurals (23.96%), 5-HMF (10.63%), total phenolics (45.5%), and acetic acid (5.54%) was observed with a marginal loss in xylose (7.51%), glucose (25.50%), and arabinose (8.56%) (Table 1). During Ca(OH)$_2$ overliming at pH 10, precipitation of furfurals and phenolics occurs, resulting in their removal during vacuum filtration of the hydrolysate. Earlier, we have reported 41.75% and 33.21% reductions in furans and phenolics from S. spontaneum hemicellulosic hydrolysates after Ca(OH)$_2$ overliming [24]. There was only a 5.5% reduction in acetic acid content after overliming. In another study, Rodrigues et al. [22] observed no change in acetic acid concentration after overliming of corn stover hemicellulosic hydrolysate.

6.2.2 ENZYMATIC HYDROLYSIS

OAFEX-treated bagasse was enzymatically hydrolyzed with commercial enzymes, i.e., 20 FPU/g of Celluclast 1.5 L and 30 IU/g of Novozym 188 at 50°C for 96 h in the presence of tween-20. Enzymatic hydrolysis after 72 h revealed 28.5 g/L glucose release with 66.51% saccharification efficiency (Figure 1). The OAFEX-treated bagasse showed a concomitant increase in sugar recovery up to 72 h followed by a decrease in glucose

TABLE 1: Sugarcane bagasse hemicellulosic hydrolysate profile after detoxification by calcium hydroxide overliming.

Constituents	Quantity (g/l)	% Reduction after overliming	Residual constituents quantity (g/l)
Xylose	12.56	7.51	11.61
Glucose	1.85	25.50	1.37
Arabinose	0.85	8.56	0.777
Acetic acid	1.47	5.54	1.38
Furfural	0.242	23.96	0.184
5-HMF	0.094	10.63	0.084
Total phenolics	2.95	45.5	1.60

The values are mean of three replicates. Standard deviation was within 10%.

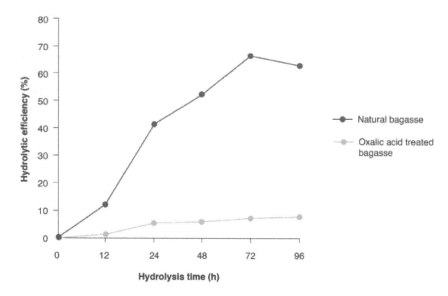

FIGURE 1: Enzymatic hydrolysis profile of oxalic acid pretreated sugarcane bagasse. The values are mean of three replicates. Standard deviation was within 10%.

concentration at 96 h of incubation. These enzyme loadings in the presence of surfactants are sufficient to hydrolyse the cellulose present in pretreated substrate.

A similar trend was reported in the sugar recovery during enzymatic hydrolysis of aqueous-ammonia-pretreated *S. spontaneum* [25]. The amounts of enzymes required for hydrolysis of pretreated raw material depend upon the pretreatment applied to the substrate and the availability of carbohydrate content in the substrate [26]. A maximum yield of sugars (482±22 mg/g, 64% hydrolytic efficiency) from acidic-hydrolyzed wheat straw (7.83% DS, acid loading 0.75%) was obtained after enzymatic hydrolysis [26]. We also observed 28.5 g/l glucose representing 66.51% efficiency after enzymatic hydrolysis of OAFEX-treated bagasse in the presence of tween-20 (2.5 g/l). Zheng et al. [27] observed that high enzyme loadings did not alter saccharification and yields. Rezende et al. [8] reported 72% cellulose conversion from consecutive acid–base pretreated SB. A 65% cellulose conversion was obtained after the enzymatic hydrolysis (1.91% w/w pretreated SB, 20 FPU/g enzyme loading, 0.05 g/g surfactant) of bagasse pretreated with dilute sulfuric acid (1.75% w/w bagasse content, 1.7% w/w H_2SO_4 loading, 150°C, and 30 min pretreatment time) [20]. Our results indicate that hemicellulose removal and the possible relocalization of lignin moieties during OAFEX treatment could yield the desired amount of sugar toward the goal of developing an intensified and simplified process for cellulose saccharification.

6.2.3 ULTRA STRUCTURAL CHARACTERIZATION OF NATIVE AND TREATED SB

SB, the biomass is naturally built through a special arrangement of crosslinked lignin with a holocellulose network, providing a superb mechanism to protect itself from microbial invasions in nature [5]. Rendering the carbohydrate fraction of the cell wall accessible is a multiscale phenomenon encompassing several orders of magnitude due to both macroscopic (compositional heterogeneity, mass transfer limitations) and microscopic barriers (holocellulose crystallinity, lignin-cellulose linkage) [28]. The pretreatment allows cellular breakdown, increasing the amenability of enzymes for sugar monomer recovery. The SEM and AFM of cell walls after acidic treatment reveal disorganization of cell wall components, which is pivotal for improved cellulase action on the carbohydrate polymer in order to yield simple sugars.

6.2.4 CELL WALL COMPOSITION

A physico-chemical analysis of the native SB cell wall revealed a composition (dry weight) of 24.67% total lignin (20.88% Klason lignin + 3.79% acid soluble lignin), 41.22% cellulose, 25.62% hemicellulose, 2.90% extractives, 8.16% moisture, and 1.5% structural ash. This composition of SB is in agreement with earlier studies [20]. The treated substrate after OA pretreatment of SB revealed 58.84% cellulose, 8.75% hemicellulose, and 24.56% total lignin (24.18% Klason lignin + 0.38% acid soluble lignin). The total lignin content in natural and OA-pretreated bagasse was determined to be almost similar (OA: 24.56%; natural: 24.67%). Klason lignin content increased in the OA-treated bagasse (24.18%) compared to the native bagasse (20.88%). However, acid-soluble lignin was removed maximally in the OA-pretreated bagasse, making the total content almost similar.

The pretreatment of corn cob with OA showed 100% removal of hemicellulose with a partial increase in lignin. Lee et al. [13] observed 6.76% xylan, 35.09% glucan, 0.12% galactan, 0.55% arabinan, and 12.09% lignin in OA-pretreated corn cob (168°C, 26 min, 30 g/l OA, 1:6 solid-to-liquid ratio) compared with the native material (27.86% xylan, 37.07% glucan, 0.61% galactan, 2.19% arabinan, and 13.92% lignin). The changes in cell wall composition of lignocellulosic material after OA pretreatment depend upon the nature of the substrate and the conditions explored during pretreatment [11-13].

6.2.5 SCANNING ELECTRON MICROSCOPY (SEM)

SEM analysis indicated the cell wall degradation, and surface properties of native, OA-pretreated, and enzymatically digested SB (Figure 2a-c). The native SB cell wall showed parallel stripes and waxes, extractives, and other deposits on the surface (Figure 2a). Obtained results are in agreement with earlier studies, where OA was reported to disrupt the structure of fibers and pith by removing the hemicellulose fraction of the cell wall, waxes, and other deposits. OAFEX leaves the overall structure disorganized, simultaneously increasing the surface area for enzymatic action [14,23].

OA-pretreated bagasse fiber showed small pores on the surface and fiber disruption, which revealed the efficacy of the pretreatment process (Figure 2b). The OAFEX-treated SB revealed disruption of cell wall after enzymatic hydrolysis. Similarly, Rezende et al. [8] observed a disrupted fiber surface and exposed parallel stripes after sulfuric acid pretreatment of SB. Our data is in agreement and indicates severe disruption and exposure of parallel stripes in OAFEX-treated SB (Figure 2b). After the enzyme treatment of SB, the analysis showed maximum disintegration and numerous holes in the cell wall, verifying the enzymatic action on

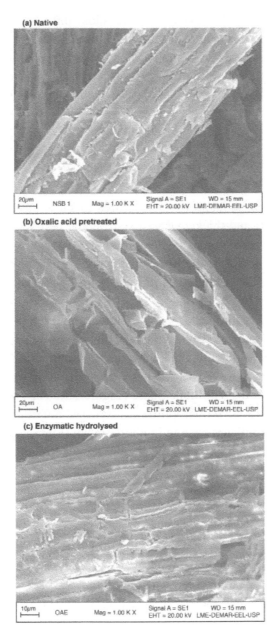

FIGURE 2: Scanning electron microscopy surface images of the sugarcane bagasse. (a) Native (b) Oxalic acid pretreated (c) Enzymatic hydrolysed. Compacted surface of cell wall showing waxes and deposits in native SB, disruption of cell wall was more evident after oxalic acid mediated pretreatment and enzymatic hydrolysis.

cellulose (Figure 2c). The exposure of cellulose through structural alteration of the bagasse is the crucial factor in hydrolysis of the remaining cellulosic fraction present in the cell wall. Similar observations have been
reported from enzymatic hydrolysis of dilute-sulfuric-acid-pretreated S.
spontaneum[24,25]. Kristensen et al. [29] also observed similar effect on
wheat straw cell walls after hydrothermal pretreatment.

6.2.6 ATOMIC FORCE MICROSCOPY (AFM)

Amplitude and phase images were captured to show the changes in secondary cell walls and thickened vascular bundle cell surfaces of native,
OAFEX-treated, and enzymatically digested SB (Figure 3a-c). The native SB showed a fibrous network of cellulose, cross-linked lignin, and
hemicellulose in the parenchyma of the primary wall (Figure 2a), which
is also shown by the AFM image, intact with a uniform surface (Figure
3a). The native SB surface was predominantly hydrophobic (95.40 nm),
as confirmed by the darker phase image (Figure 3a). OA specifically disrupts hemicellulose, retaining cellulose and lignin together. The OAFEX-
treated SB clearly showed the non-homogeneous phase and globular surface deposition in the amplitude phase (Figure 3b). The AFM tip showed
increased affinity toward hydrophilic regions (192 nm) that appear light in
color due to a significant change in phase image (Figure 3b). The surface
of OA-treated SB is non-uniform, encompassing irregularly shaped hydrophilic deposits, probably due to the exposure of cellulose. Chundawat
et al. [28] observed similar patterns in corn stover after pretreatment with
ammonium hydroxide.

The presence of globular and irregular shapes (20–95 nm in diameter)
can be characteristic of lignin deposits in the cell wall (Figure 3b). This
interpretation is in accordance with the SEM analysis, where re-localization of lignin is apparent, linking with cellulose lamellae/ agglomerates
(Figure 2b). Lignin re-localization is an important feature that may lead to
enhanced enzymatic hydrolysis. OA markedly disrupts hemicellulose and
simultaneously re-localizes lignin moieties, aiding the increased exposure
of cellulose to cellulases [29,30].

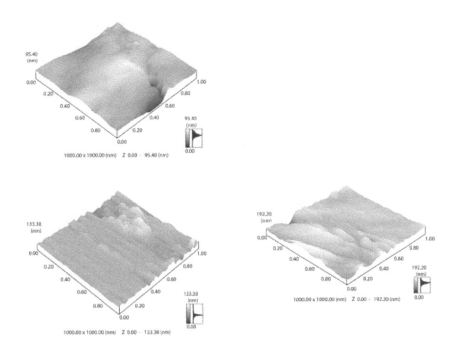

FIGURE 3: Atomic force microscopy (AFM) amplitude images. (a)Native sugarcane bagasse (b) Oxalic acid-pretreated bagasse (c) Enzyme hydrolysed bagasse. AFM scan revealed the cross linked cellulose + hemicellulose network in native SB. After OA pretreatment, non homogenous surface appeared with globular surface deposition (characteristic of lignin). AFM tip indicated affinity towards hydrophilic areas (192 nm) revealing the breakdown of cellulose.

The higher-resolution imaging of AFM reveals maximum non-uniformity in the cellulose lamellae, which can be interpreted as the complete disruption of cellulose aggregates into glucose (Figure 3c). The AFM tip showed high affinity toward hydrophilic areas (192 nm) appearing as changes in phase, and the light color shows the effective accessibility of cellulose to cellulases. Igarashi et al. [31] reported that the real-time visualization of cellulase from *Trichoderma reesei* cellobiohydrolase I action on crystalline cellulose resulted in marked cellulase affinity toward cellulose.

6.2.7 X-RAY DIFFRACTION (XRD)

XRD analysis revealed the increasing order of crystallinity index (CrI) in OAFEX-treated bagasse (52.56%) and enzyme-digested bagasse (55.65%) compared to native bagasse (45.61%). Figure 4 shows the XRD spectra of native, OAFEX-treated, and enzyme-digested bagasse. In general, after pretreatment of SB, CrI increases with the decrease of amorphous regions in the substrate along with other constituents [6]. The CrI of OAFEX-treated bagasse increases mainly due to removal of hemicellulose in conjunction with re-localisation of lignin and partial disruption of cellulose. Rezende et al. [8] reported that the CrI of consequentially acid–base-pretreated SB increased with a parallel increase in the amount of cellulose in the substrate. Velmurugan and Muthukumar [7] observed a high CrI in sono-assisted alkali-pretreated SB (66%) compared with natural bagasse (50%). The increase in CrI of OAFEX-treated bagasse and enzyme-digested bagasse was due to the depolymerization of hemicellulose and cellulose into their monomeric constituents. Cellulase enzyme cocktails break down the amorphous cellulose, thus increasing the CrI over OAFEX-treated bagasse. However, cellulose crystallinity is not considered to be a principal factor that determines biomass recalcitrance [8].

6.2.8 RAMAN SPECTROSCOPY

Raman spectroscopic analysis revealed a gradual reduction in the band intensity (< 1500 cm−1) in OAFEX-treated and enzymatically hydrolyzed SB compared to the native bagasse (Figure 5). The molecular disarrangement and displacement in the hemicellulosic backbone during pretreatment followed by enzymatic cleavage of the β–1–4–glycosidic linkage of cellulose may have caused this gradual reduction in overall band intensities. Previous studies that performed Raman spectroscopic analysis of cellulose and hemicellulose showed that the changes in spectra were dominated by contributions from cellulose at intensities below 1500 cm^{-1}[32]. The ratio of the bands at 1172 cm^{-1} and 1204 cm^{-1} reflects the orientation of the cellulose relative to the electric vector (polarization) of the laser [33].

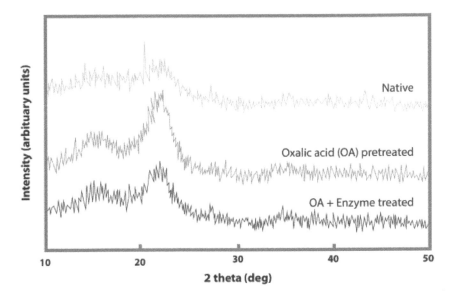

FIGURE 4: X-ray diffraction pattern of native, oxalic acid pretreated and enzyme hydrolysed SB. Crystallinity Index (CrI) was found to be increased in OA pretreated SB and enzymatic hydrolysed SB showing the effect of OAFEX and enzymatic degradation.

In Raman analysis, the electric field vector has a component along the cellulose axis causing the intensification of the band at 1172 cm⁻¹, and considered to obtain the spectral analysis of native, OAFEX-treated, and enzyme-digested SB. Figure 5 shows cellulose/ hemicellulose peaks at 1088 and 1371 cm⁻¹. It is evident here that OA pretreatment and enzymatic hydrolysis act on hemicellulose and cellulose respectively. In addition, the main signature of lignin in all three samples is strong band lines at 1603 and 1630 cm⁻¹ due to stretching of the asymmetric aryl ring in lignin [33,34]. This may be due to the re-localization of lignin moieties during OA pretreatment of SB at a high temperature. Ooi et al. [34] reported that Raman spectra of native and NaOH-treated Kenaf fibers showed a reduction in intensity of lignin bands in the region around 1750 cm⁻¹ assigned to acetyl groups with the carbonyl, C=O group. No significant changes were

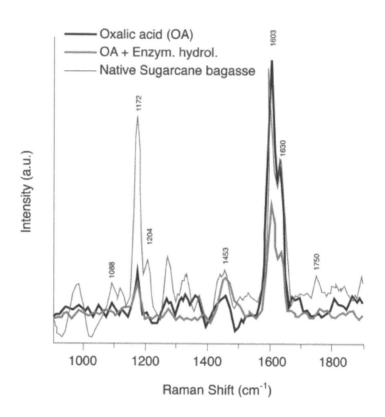

FIGURE 5: Raman spectra of native sugarcane bagasse (thin grey line), pre-treated bagasse with oxalic acid (thick black line) and oxalic acid/enzymatic hydrolysis pre-treated bagasse (thick grey line). A gradual reduction in band intensities was recorded in OAFEX and enzymatic hydrolysis revealing the displacement in hemicellulose backbone and enzymatic cleavage of β–1–4–glycosidic linkage in cellulose.

observed in the intensities of lignin bands at 1750 cm⁻¹, maybe due to the fact that OA and enzymatic hydrolysis specifically act on hemicellulose and cellulose respectively.

6.2.9 FOURIER TRANSFORM INFRARED SPECTROSCOPY (FTIR)

FTIR spectroscopy was carried out to investigate the changes in hemicellulose and cellulose structure during OAFEX treatment and enzymatic hydrolysis. FTIR spectra of the native SB show a band at 900 cm^{-1} representing the β-(1–4) glycosidic linkages of cellulose (Figure 6a-d). The frequency range between 1200–1000 cm^{-1} has a large contribution of hemicellulose and cellulose with maxima at 1037 cm^{-1} due to C-O stretching mode and 1164 cm^{-1} due to the asymmetrical stretching C-O-C [35,36]. The absorption at 1247 cm^{-1} shows due to C–O stretching and is a feature of the hemicelluloses, as well as of lignin [37]. There is divergence about the region at 1316 cm^{-1} (Figure 6b), which can be attributed to vibration of the CH2 group of cellulose [35]. Pandey attributes it to the C-O syringyl ring in lignin [38].

The absorption around 1463 cm^{-1} refers to CH$_2$ and CH$_3$ deformation of lignin, while 1606 cm–1 is related to C=C stretching of the aromatic ring (lignin) and C=O stretching (Figure 6c). The absorption around 1515 cm^{-1} is associated with C=C aromatic skeletal vibration [35,38]. The absorption in 1733 cm^{-1} is attributed to a C=O unconjugated stretching of hemicelluloses but also with the contribution of lignin [35,38]. The small peaks at 2850 cm^{-1} and 2918 cm^{-1} come from CH$_2$ and CH symmetric and asymmetric stretching respectively (Figure 6d). Both are characteristic of cellulose [36]. The obtained data agree that the range 3800–3000 cm–1 comprises bands related to the crystalline structure of cellulose [35]. The region is of great importance and is related to the sum of the valence vibrations of H-bonded OH and intramolecular and intermolecular hydrogen bonds. In the range 1300–1000 cm^{-1}, the appearance of two peaks at 1033 cm^{-1} and 1058 cm^{-1} were observed in OAFEX-treated bagasse and enzyme-digested cellulignin spectra respectively (Figure 6b). This indicates penetration of OA in the amorphous region of the biomass and degrading hemicellulose.

Further, the removal of the hemicelluloses is clearly seen by comparing the band around 1247 cm^{-1} with the abrupt decrease of the same band in the OAFEX-treated bagasse (Figure 6b). In the range 1800–1400 cm^{-1}, we did not observe changes except around 1733 cm^{-1}, which indicates chemical changes in hemicellulose and/or lignin (Figure 6c). Nevertheless, no apparent changes were observed in the lignin-characteristic bands around 1606 cm^{-1}, 1515 cm^{-1} and 1463 cm^{-1}. Therefore, it is reasonable to state that the hemicellulose was degraded by the action of the OA, which appears to be less stable than lignin. These observations could reflect that

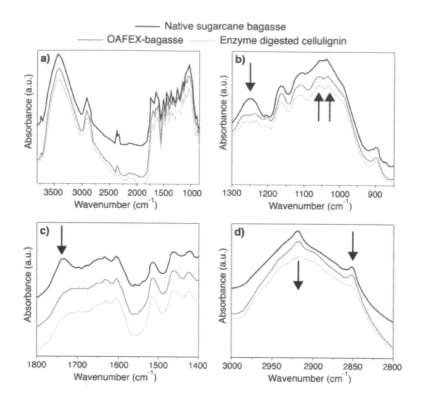

FIGURE 6: FTIR spectra of Native sugarcane bagasse (black line), Oxalic acid fiber expansion (OAFEX) (dark grey line) and Enzyme digested cellulignin (light grey line). 7b), 7c) and 7d) selected regions of Figure 7a. The arrows indicate changes observed through the pretreatment. Hemicellulose and cellulose maxima appeared between frequency ranges of 1000–1200 cm⁻¹ with maxima at 1037 cm⁻¹ due to C-O stretching. The absorption around 1463 cm⁻¹ refers to CH_2 and CH_3 indicating the deformation of lignin while 1606 cm⁻¹ is related to C=C stretching of the aromatic ring (lignin) and C=O stretching.

the lignin composed of percentages of p-hydroxyphenyl (H), guaiacyl (G), and syringyl (S) units is highly condensed and very resistant to degradation. Further, the 3000–2800 cm⁻¹ range shows a decrease of two local maxima mainly in the enzyme-digested cellulignin, indicating the cellulose hydrolysis (Figure 6d).

FTIR spectroscopy, together with the deconvolution technique, provides distinct positions of the bands in this region. Gaussian distribution of the modes in the deconvolution process showed the three bands (Figure 7a-c). The band range from 3310 to 3228 cm⁻¹ refers to the intermolecular hydrogen bond O(6)H....O(3) (Figure 7), 3375 to 3335 cm⁻¹ belongs to the intramolecular hydrogen bond O(3)H....O(5), and 3460 to 3410 cm⁻¹ is related to the intramolecular hydrogen bond O(2)H....O(6) (Figure 7). The band around 3585 cm⁻¹ (band 1 in Figure 7) has been reported as the contribution of free hydroxyl [39].

The deconvolution FTIR spectra reveal three bands of the crystalline structure of cellulose (Figure 7). Band 3 of intermolecular hydrogen bonds has shifted to a higher wave number (3220 cm⁻¹; 3239 cm⁻¹; 3246 cm⁻¹), revealing the depolymerization of crystalline cellulose (Figure 7b). Band 2 of intramolecular hydrogen bonds has also shifted to higher frequencies

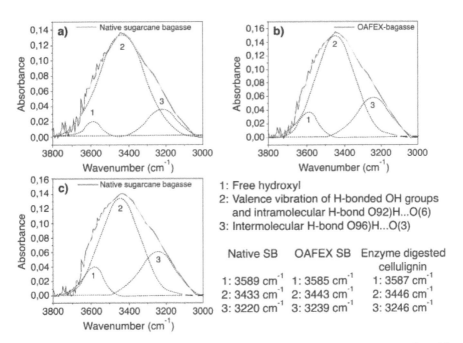

FIGURE 7: Deconvoluted FTIR spectra of a) Native sugarcane bagasse, b) Oxalic acid fiber expansion (OAFEX) and c) Enzyme digested cellulignin showing: the free hydroxyl (band 1), Valence vibration of H-bonded OH groups and intramolecular H-bond (band 2) and Intermolecular H-bond (band 3).

(3433 cm⁻¹; 3443 cm⁻¹; 3446 cm⁻¹) indicating the formation of intramolecular hydrogen bonds when OA penetrated the crystalline structure of cellulose (Figure 7c). This shift is evidence of an energy change in the internal interactions of cellulose [40,41]. The disruption of the structure of plant vegetal fibers and the removal of hemicellulose and/or lignin from the polymer matrix can also configure a closer relationship between the cellulose chains. Another indication of this behavior is the increase in the width and asymmetry of the curves for the OAFEX-treated bagasse.

6.2.10 NEAR INFRARED SPECTROSCOPY (NIR)

FT-NIR spectra of native, OAFEX-treated, and enzyme-digested SB ranged from 7200 to 4000 cm⁻¹ (Figure 8). The spectral region from 6000 to 5920 cm−1 is assigned to the first overtone C-H stretching vibration of aromatics. This could be due to the re-localization of lignin moieties during OA-mediated pretreatment at a high temperature [42,43]. In regard to the hemicellulose structural change, the first overtone C-H stretch around 5800 cm⁻¹ is responsible for its variation [42,44]. This reveals that the absorbance of the hemicellulose decreases with diminishing hemicellulosic content in SB. This drastic change in absorbance in hemicellulose is due to the severe pretreatment given to the SB [42].

The stressed regions (grey) around 7000 cm⁻¹ correspond to amorphous cellulose (Figure 8). The spectral band around 6290 ±20 cm⁻¹ is attributed to crystalline cellulose (CII) and the presence of polysaccharides is shown at 5400 to 4000 cm⁻¹ [42,44]. Additionally, we observed changes in the polysaccharide content in the regions around 6300, 5208, 4813, 4285, and 4405 cm⁻¹. The amorphous region of polysaccharide around 7000 cm⁻¹ had a small increase, suggesting the re-localization of lignin units [42].

6.2.11 ETHANOL FERMENTATION

Detoxified hemicellulosic hydrolysate and enzymatic hydrolysates were used for ethanol production using yeasts *C. shehatae* UFMG52.2 and *S. cerevisiae* 174 under submerged culture cultivation. The microorganisms

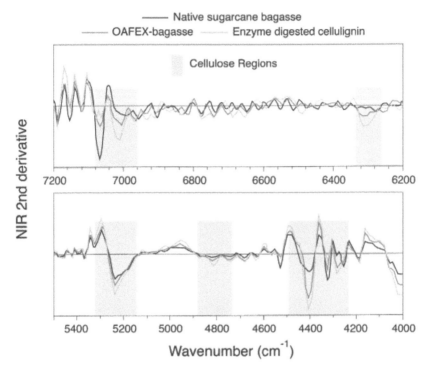

FIGURE 8: Second derivative of the NIR spectra (7200 to 4000 cm^{-1}) of native sugarcane bagasse (black line), pre-treated bagasse with oxalic acid (dark grey line) and enzymatic hydrolyzed bagasse (light grey line). In the OAFEX-SB, spectral region 6000 to 5920 cm^{-1} is assigned to the C-H stretching vibration of aromatics due to the re-localization of lignin. Changes in the polysaccharide content after OAFEX and enzyme hydrolysis at regions around 6300, 5208, 4813, 4285, and 4405 cm^{-1}.

selected for bioconversion of ethanol have been established for the fermentation of xylose and glucose sugars for ethanol production [17,19].

6.2.12 FERMENTATION OF ACID HYDROLYSATE

The fermentation profile of detoxified SB acid hydrolysate from *C. shehatae* UFMG52.2 and *S. cerevisiae* 174 in batch culture is shown in Figure 9a, b. The microorganism *C. shehatae* reached maximum ethanol produc-

(a)

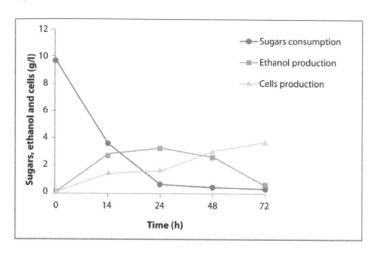

(b)

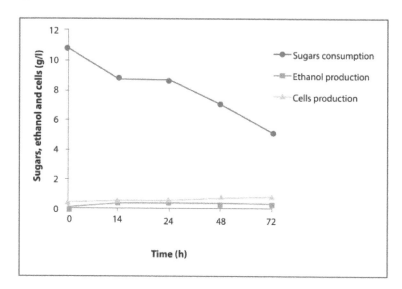

FIGURE 9: The time course of growth, sugar utilization and ethanol production using detoxified oxalic acid hydrolysate (fermentable pH of the hydrolysate was adjusted 5.5) by (a) *C. shehatae* UFMG HM52.2 at 30°C (b) *S. cerevisiae* at 30°C. The values are mean of three replicates. Standard deviation was within 10%.

tion (3.20g/l) with a yield of 0.353 g/g and productivity of 0.133 g/l/h from detoxified hemicellulosic acid hydrolysate after 24 h, and declined afterwards (Figure 9a). The biomass continued to increase even after 24h to the completion of the fermentation cycle (72 h), and yielded 0.385 g/g with productivity of 0.0496 g/l/h (Table 2). The detoxified hemicellulose hydrolysate did not show satisfactory ethanol production (Figure 9b) by *S. cerevisiae*, which suggests that the microorganism was unable to utilize the abundance of xylose sugar in the hydrolysate. After 72 h of incubation, a

TABLE 2: Kinetic parameters for ethanol production from detoxified oxalic acid hydrolysate and enzymatic hydrolysates by *Candida shehatae* UFMG HM52.2 and *Saccharomyces cerevisiae* 174.

Parameters	Hemicellulose hydrolysate #		Enzyme hydrolysate ##	
	C. shehatae	*S. cerevisiae*	*C. shehatae*	*S. cerevisiae*
Initial sugars (g_s/l)	9.61	10.25	21.03	18.4
Residual sugars (g_s/l)	0.56	8.14	3.96	0.87
Ethanol (g_p/l)	3.20	0.52	4.83	6.6
Ethanol yield (g_p/g_s)	0.353	0.246	0.282	0.46
Ethanol productivity (g_p/l/h)	0.133	0.021	0.201	0.47
Biomass (g_x/l)	3.57	0.85	6.32	4.03
Biomass yield (g_x/g_s)	0.385	0.179	0.302	0.22
Biomass productivity (g_x/l/h)	0.0496	0.011	0.0877	0.055

During the fermentation of hemicellulosic hydrolysate by *C. shehatae* UFMG HM52.2 and *S. cerevisiae* 174, maximum ethanol was produced after 24 hrs, so ethanol productivity was calculated in both the cases considering 24 hrs. After 24 hrs of incubation, a concomitant downfall in ethanol production was observed with a regular increase in biomass. However, to calculate biomass productivities, incubation time (72 hrs) was considered.

During the fermentation of enzyme hydrolysate by *C. shehatae* UFMG HM52.2 and *S. cerevisiae* 174, maximum ethanol was produced after 24 hrs and 14 hrs respectively. Therefore ethanol productivities were calculated considering these incubation times. However biomass was found consistently increasing until the completion of fermentation (72 hrs), so biomass productivities were calculated considering 72 hrs of incubation. The values are mean of three replicates. Standard deviation was within 10%.

biomass yield of 0.179 g/g and productivity of 0.011 g/l/h were observed (Table 2).

6.2.13 FERMENTATION OF ENZYME HYDROLYSATE

When the enzymatic hydrolysate was fermented with *C. shehatae* UFMG52.2, maximum ethanol production (4.83 g/l) was found with a yield of 0.282 g/g and productivity of 0.201 g/l/h after 24 h (Figure 10a; Table 2). However, regular growth of microorganisms was observed until the sugar was consumed after 72 h. The maximum biomass production (6.32 g/l) was obtained with a yield of 0.302 g/g and productivity of 0.877 g/l/h. Fermentation of enzymatic hydrolysate with *S. cerevisiae* showed maximum ethanol production (6.6 g/l) after 24 h, with a yield of 0.46 g/g and productivity of 0.47 g/l/h after 14 h of incubation (Figure 10b). The growth in biomass remained unchanged until 72 h of incubation (Figure 10b; Table 2).

C. shehatae showed a greater ethanol yield (0.353 g/g) from the acid hydrolysate than from the enzymatic hydrolysate (0.282 g/g), indicating a microbial preference for xylose over glucose as a source of carbon in the fermentation reaction. However, *C. shehatae* is sensitive to fermentation inhibitors present in the hydrolysate. Sreenath et al. [45] showed ethanol production of 5 g/l with a yield of 0.25 g/g by *C. shehatae* FPL 702 from alfalfa hydrolysate. The low ethanol yield of this yeast was probably due to inhibition by pectic acid, organic acids, and hemicellulose-derived inhibitors present in the sugar solution [45]. The hemicellulosic hydrolysates contain mainly xylose, while enzymatic hydrolysates contain only glucose. When *S. cerevisiae* was grown on hemicellulosic hydrolysate, abysmal ethanol production and growth were recorded (Table 2). *S. cerevisiae* does not use xylose present in the hemicellulosic hydrolysate, but rather relies upon glucose, which could be why the poor ethanol production was observed (Figure 9b). Our earlier studies showed that S. spontaneum acid hydrolysate after overliming fermented with *S. cerevisiae* VS3 in batch conditions produced 1.40±0.07 g/l ethanol with 0.20±0.016 g/g yield after 36 h of fermentation [24].

(a)

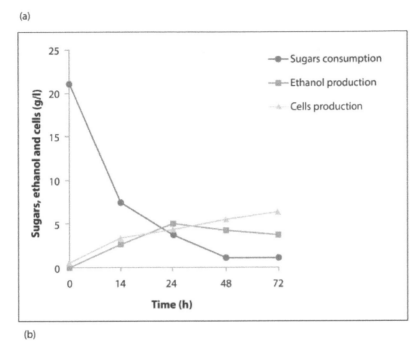

(b)

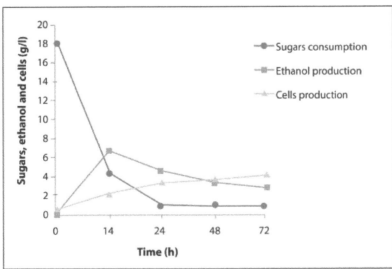

FIGURE 10: The time course of growth, sugar utilization and ethanol production using enzyme hydrolysate (fermentable pH of the hydrolysate was adjusted 5.5) by (a) *C. shehatae* UFMG HM52.2 at 30°C (b) *S. cerevisiae* at 30°C. The values are mean of three replicates. Standard deviation was within 10%.

C. shehatae showed a similar growth pattern in both the hydrolysates. In both cases, more than 80% of the sugars were utilized within 24 h of incubation promoting faster growth of microorganism. Similar patterns of biomass growth were observed by Abbi et al. [46], and Chandel et al. [17] reported a regular increase in biomass of *C. shehatae* NCIM 3501 after the exhaustion of xylose in 24 h with the utilization of ethanol as a carbon source for the metabolic growth. Sánchez et al. [47] observed ethanol production (4.5 g/L) from *C. shehatae* CBS4410 utilizing Paja brava acid hydrolysate (19.8 g/L xylose, 2.5 g/L glucose) obtained under hydrolysis conditions (180°C, 5 min, 0.5% w/w H2SO4). Sreenath and Jeffries [48] reported that *C. shehatae* FPL-Y-049 utilized all sugars present in wood hemicellulose hydrolysate except arabinose and produced 34 g/L of ethanol with a yield of 0.46 g/g.

The fermentation of enzymatic hydrolysate (18.4 g/l glucose) by *S. cerevisiae* produced maximum ethanol (6.6 g/L, yield 0.46 g/g) after 14 h of incubation under conditions similar to those employed for hemicellulose hydrolysate fermentation (Figure 10b). It is interesting to note that both yeast strains isolated from Brazilian biodiversity presented a higher growth rate with a shorter lag phase and a prolonged exponential phase. The current growth pattern of both microorganisms supports maximal conversion of sugars into ethanol-an impressive trait for achieving higher ethanol productivity. These characteristics could be beneficial in biorefineries by saving time and reducing the cost of operation and energy.

The maximum ethanol productivity (0.47 g/l/h) was obtained from *S. cerevisiae*-mediated fermentation of OAFEX-treated SB enzymatic hydrolysate, another impressive feature of the *S. cerevisiae* strain used in this study (Table 2). Previously, a natural isolate from spent sulphite liquor, *S. cerevisiae* ATCC 96581, showed maximum ethanol production (7.4 g/l, yield 0.28 g/g and productivity 0.37 g/l/h) from detoxified SB hydrolysate containing 26.0 g/l total sugars [19]. Unlike acid hydrolysates, enzymatic hydrolysate lacks fermentative inhibitors, which eliminates the detoxification step. During the fermentation of enzymatic hydrolysate, *S. cerevisiae* showed less biomass production (4.03 g/l, yield 0.22 g/g) compared to *C. shehatae* (6.32 g/l, yield 0.302 g/g) after 72 h of incubation, suggesting that *C. shehatae* prefers to metabolize a carbon source for ethanol production rather than produce cellular buildup. However, the characteristic of selecting ethanol as a carbon source after the exhaustion of sugars was also observed in *S. cerevisiae* 174.

6.3 CONCLUSION

Plant cell walls are complex, rigid, and recalcitrant in nature. Pretreatment is an inevitable process to break down this carbohydrate skeleton and increase the accessibility of cellulose. Our studies indicated that OA-mediated hemicellulose degradation of SB is an effective pretreatment strategy that ameliorates the enzymatic hydrolysis of the cellulosic fraction into glucose. The microscopic and spectroscopic techniques used in this work warranted in-depth structural investigation of chemical changes at the molecular level during pretreatment and enzymatic digestion. OAFEX pretreatment significantly removed hemicellulose, causing lignin re-localization, which eventually showed efficient enzymatic action toward the depolymerization of cellulose into glucose (66.51% hydrolytic efficiency). Detoxified hemicellulosic hydrolysate, when fermented by *C. shehatae* UFMG52.2 and *S. cerevisiae* 174, showed ethanol production of 3.20 g/L (yield 0.353 g/g) and 0.52 g/L (yield 0.246 g/g) respectively. Enzymatic hydrolysate after fermentation with *C. shehatae* UFMG52.2 and *S. cerevisiae* 174 produced 4.83 g/L (yield 0.282 g/g) and 6.6 g/L (yield 0.46 g/g) respectively. Both microorganisms revealed high substrate consumption including energy and time savings that could have a major impact in biorefineries.

6.4 MATERIAL AND METHODS

6.4.1 *PREPARATION OF RAW SUBSTRATE*

The raw substrate, sugarcane bagasse, was acquired from Usina Vale do Rosário (Morro Agudo, S.P, Brazil). In preliminary processing, the SB was air-dried, and knife milled (Marconi Equipamentos, Model No. MA 680, Piracicaba-S.P, Brazil) to pass through with 20-mesh sieve. The finely milled bagasse was washed under running tap water to remove the dust, and dried at 45°C for further experiments.

6.4.2 OXALIC ACID FIBER EXPANSION PRETREATMENT OF SUGARCANE BAGASSE

The pretreatment of the SB with dilute oxalic acid (OA) (3.5% w.v−1) was carried out at 160°C for 20 min as described by Scordia et al. [12]. Briefly, SB (10 g d.wt.) and 100 ml (3.5% w.v−1) aqueous OA solution (10% total solids) were loaded into a 200 ml stainless-steel container (19x9.7 cm), tightly sealed and immersed in an oil bath provided with electrical heating at 160°C. The container remains resident for 5 min to reach the tempera-ture of 160°C. The hydrolysis was stopped through immersing the con-tainer into running water in due course.

After hydrolysis, the hemicellulosic hydrolysate was quantitatively separated by vacuum filtration from the pretreated solids, hereinafter re-ferred to as cellulignin. The cellulignin was thoroughly washed with de-ionized water and dried in oven at 45°C for 72 h, and subsequently used for enzymatic hydrolysis.

6.4.3 DETOXIFICATION OF OXALIC ACID HYDROLYSATE

OA hydrolysate was detoxified using overliming with the addition of dried calcium oxide under constant stirring until the pH reached 10.5 ± 0.05. The fermentation inhibitors were allowed to precipitate for additional 1 h by stirring. The slurry was then subjected to vacuum filtration using What-man filter paper #1 to remove the precipitates. The pH of clear filtrate was adjusted to 6.00 ± 0.05 with 6 N H_2SO_4 and again vacuum filtered to remove traces of salt precipitates.

6.4.4 ENZYMATIC HYDROLYSIS

OAFEX-bagasse was enzymatically hydrolysed to depolymerise carbo-hydrates into simpler sugars. The OAFEX-bagasse (2 g d.wt) was pre-incubated in 40 ml of sodium citrate buffer (50 mM, pH 4.8) in 150 ml Erlenmeyer flask for 1 h at room temperature. The microbial growth was restricted by adding sodium azide (0.005%) during enzymatic hydrolysis.

Soaked OAFEX-bagasse was supplemented with different cellulase loadings i.e. 20 FPU/g of the dry substrate from Celluclast 1.5 L, and 30 IU/g of β-glucosidase from Novozym 188 (Sigma-Aldrich, USA). The reaction mixture was supplemented with non-ionic surfactant (2.5 g/l, Tween–20; polyoxyethylene sorbitan monolaurate). The enzymatic hydrolysis was performed at 50°C at 150 rpm in incubator shaker (Innova 4000 Incubator Shaker, New Brunswick Scientific, Enfield, CT, USA) for 96 hrs. The samples (0.5 ml) were collected periodically (24 h), centrifuged at 5000 rpm at room temperature. The supernatant was analyzed for total reducing sugars using DNS method [49]. The extent of hydrolysis (after OA pretreatment and enzymatic hydrolysis) was calculated as follows:

Hydrolysis (%) = reducing sugar concentration obtained × 100/Total carbohydrate content (TCC) in sugarcane bagasse.

6.4.5 FERMENTATION

6.4.5.1 MICROORGANISMS AND INOCULUM PREPARATION

Microorganisms *Candida shehatae* UFMG HM 52.2 and *Saccharomyces cerevisiae* 174 were used for the fermentation of the hemicellulose hydrolysates and enzymatic hydrolysates. The microorganism *C. shehatae* UFMG HM 52.2 was isolated from a rotting-wood sample in an Atlantic Rain Forest site (Bello & Kerida Ecological Reserve) situated in the city of Nova Friburgo, Rio de Janeiro state as described by Cadete et al. [50]. *S. cerevisiae* 174 was isolated from the Atibaia river, São Paulo State, Brazil. Both microorganisms were maintened on YPD plates and stored at 4°C.

C. shehatae UFMG HM 52.2 was grown in 250 ml Erlenmeyer flasks containing 50 ml of seed medium (30 g/L of sugars (1:1 xylose and glucose), 20 g/l peptone, and 10 g/l yeast extract) in an orbital incubator shaker at 200 rpm at 30°C. Microorganism *S. cerevisiae* was grown in 250 ml Erlenmeyer flasks containing 50 ml of seed medium (30 g/l glucose, 5 g/l peptone, 3 g/l yeast extract, and 0.25 g/l Di-ammonium hydrogen phosphate) in an orbital incubator shaker at 100 rpm at 30°C for 24 h. The cultured microorganisms (*C. shehatae* and *S. cerevisiae*) were centrifuged and prepared for inoculum corresponding to 0.5 g/l cells (d. wt). The prepared inoculums were aseptically transferred into detoxified hemicellulosic and enzymatic hydrolysates (50 ml) supplemented with medium ingredients.

6.4.5.2 ETHANOL FERMENTATION

The fermentation medium (50 ml) of microorganism *C. shehatae* UFMG HM 52.2 was composed of the hydrolysates (detoxified and enzymatic) supplemented with (g/l): yeast extract (3.0), malt extract (3.0), and ammonium sulfate (5.0) at pH 5.5 [51]. In another set of fermentation, microorganism *S. cerevisiae* 174 was grown in the hydrolysates (detoxified and enzymatic) supplemented with (g/l): yeast extract (1.0), peptone (1.0), di-ammonium hydrogen phosphate (1.0), di-potassium hydrogen phosphate (1.0), magnesium sulfate and manganese sulfate (0.5) at pH 5.5 [24]. The hydrolysates were sterilized at 120 °C for 15 min before the use and the medium components were added aseptically prior to addition of inoculum. Fermentation reactions were setup in an orbital incubator shaker (Innova 4000 Incubator Shaker, New Brunswick Scientific, Enfield, CT, USA) at 200 rpm (*C. shehatae* UFMG HM 52.2), and 100 rpm (*S. cerevisiae* 174) for 72 h. Samples (1.0 ml) were collected periodically to determine the production of ethanol, residual sugars and formation of biomass in fermentation broth.

6.4.6 ANALYSES

The chemical composition pulverized sugarcane bagasse (native), OA-FEX-SB, EH-SB was determined according to the method of Gouveia et al. [52]. Glucose, xylose, arabinose, and acetic acid, concentrations were determined by HPLC (Waters) using Biorad Aminex HPX–87H column at 45°C equipped with refraction index detector. The mobile phase was constituted with sulfuric acid 0.01 N at 0.6 mL/min flow rate as eluent. Furfural and HMF concentrations were also determined by HPLC equipped with Hewlett-Packard RP18 column and UV–VIS detector (2489) (276 nm) at 25°C. Samples were eluted by acetonitrile/water (1:8) supplemented with

1% acetic acid (volume basis) as the eluent at a flow rate of 0.8 ml/min. Total phenolic compounds in hydrolysates were estimated calorimetrically using Folin–Ciocalteu method [53]. Ethanol production was analyzed by HPLC (Waters) using a refraction index detector (2414) and a Biorad Aminex HPX-87H column at 45°C. The growth in biomass *C. shehatae* UFMG 52.2 and *S. cerevisiae* 174 was determined at 600nm using spectrophotometer (Beckman DU640B, USA). The measured absorbance was correlated with the cell concentrations (g/l) following the calibration equation:

Y=2.0029x + 0.0056 (for *C. shehatae* UFMG 52.2)

Y=1.0804x + 0.006 (for *S. cerevisiae* 174)

Pretreatment, enzymatic hydrolysis and fermentation experiments were carried out in triplicates. The values are mean of three replicates.

6.4.7 MULTI-SCALE VISUAL ANALYSIS

6.4.7.1 SCANNING ELECTRON MICROSCOPE (SEM)

The SEM analysis of native, OAFEX and enzymatically hydrolysed SB was performed as described by Kristensen et al. [29]. Briefly, native, OA-FEX-pretreated and enzymatically hydrolysed SB distributed on a 12 mm glass cover slip coated with poly-L-lysine (Sigma Diagnostics, S.P. Brazil). The dried sections were mounted on aluminum stubs, sputter-coated (JEOL JFC–1600) with a gold layer, and used for scanning. The prepared samples were scanned and imaged using Hitachi S520 scanning electron microscope (Hitachi, Tokyo, Japan).

6.4.7.2 ATOMIC FORCE MICROSCOPY (AFM)

The AFM analysis of native, OA pretreated and enzymatically hydrolysed SB was performed as described by Kristensen et al. [29]. All AFM measurements were made with a multi-mode scanning probe microscope with a Nanoscope IIIa controller (Shimadzu SPM-9600 Deluxe, Japan). The images were acquired in tapping mode with etched silicon probe (Nanoworld Point Probe NCHR–10, 320 kHz, 42 N/m). An auto-tuning resonance frequency range of approximately 286–305 kHz with a scan rate of 0.5 Hz and sweep range of 10 kHz was used. The drive amplitude and amplitude set-point were adjusted during measurements to minimize scanning artifacts. Height, amplitude and phase images were captured simultaneously. Scan size varied from 500 nm to 5.0 μm, usually 1 μm.

Samples were fixed on metal discs with double-sided adhesive tape, and the images were measured in air. Images were collected from a minimum of 10 different fibers for each treatment with representative images displayed in the present work. The external vibration noise was eliminated using active vibration-damping table. AFM images were recorded (512 dpi), analyzed and processed (illumination and plane fitting) by the accompanying software.

6.4.7.3 X-RAY DIFFRACTION (XRD)

The crystalline nature of native, OA-treated and enzyme digested SB was analyzed by using a Scifert Isodebye Flex 3003 X-ray diffractometer (Germany). The crystallinity was analysed by adjusting the diffractometer set at 40 KV, 30 mA; radiation was Cu Kα (λ = 1.54 Å). Samples were scanned over the range of 100 <2θ <500 with a step size of 0.05° and the crystallinity index (CrI) were determined using the empirical method described by Segal et al. [54]:

$$CrI = \frac{I\ crysalline - I\ amorphous}{I\ crysalline} \; X \; 100\%$$

Where, I crysalline = intensity at 21°C and I amorphous = intensity at 18.8°C.

6.4.7.4 RAMAN SPECTROSCOPY

The Raman spectra of native, OA pretreated and enzyme digested SB were recorded via microRaman (T64000 Horiba Jobin Ivon, France) using the 488nm line of an argon laser, 1800 g/mm grating with a 50x objective. The samples were passed through the laser power of ~2 mW, and the scattered light were detected by a CCD system cooled with liquid nitrogen at the temperature of −130°C. To maximize the Raman signals, we have used the fact that the Raman intensity of the cellulose is dependent of the laser light polarization. Cellulose mediated polarization of laser light is dependent on the position of sample at the microscopic stage. The spectra are scaled to the cellulose peak at 1172 cm⁻¹, baseline was subtracted and smoothed with adjacent-averaging with 30 points.

6.4.7.4 FOURIER TRANSFORM INFRARED SPECTROSCOPY (FTIR)

FTIR spectroscopic analysis was performed to detect the changes in functional groups after OA pretreated and enzyme digested SB in comparison to native SB. Samples were put to mill in agate cups for 1.5 hrs with 400 rpm (RETSCH® PM 400) followed by passing through with an abronzinox sieve of 100 mesh size with an aperture of 150 μm. The pellets were prepared by mixing of 300 mg of spectroscopic grade KBr with 3 mg of sample in an agate mortar. Each sample was then submitted to 10 tons for 3 min in a hydraulic press (SPECAC 25T, ATLASTM). The spectra were collected in a VERTEX 70 spectrometer (Bruker Optics, Germany) and submitted to 4 cm⁻¹ of resolution and 72 scans per sample.

6.4.7.5 FOURIER TRANSFORM-NEAR INFRA-RED SPECTROSCOPY (FT-NIR)

The FT-NIR spectroscopy of native, OA pretreated and enzyme digested SB samples was performed with a spectrometer FT-NIR MPA (multi-purpose analyser) from Bruker Optics, Germany. The measurements have used to diffuse reflectance which were analysed via an integrating macro sample sphere, the diameter of measured area was 15 mm, 32 scans per sample were performed with a resolution of 4 cm^{-1} covering a range from 13000 to 3500 cm^{-1}. Second derivative spectra were calculated with 21 smoothing points after unit vector normalization. All calculations were conducted with OPUS 6.5 version software.

REFERENCES

1. Kerr RA: Energy supplies. Peak oil production may already be here. Science 2011, 331:1510-1511.
2. Goldemberg J: Ethanol for a sustainable energy future. Science 2007, 315:808-810.
3. Huber G, Dale BE: Grassoline at the Pump. Sci Amer 2009, 52-59. July
4. Chandel AK, Silva SS, Carvalho W, Singh OV: Sugarcane bagasse and leaves: Foreseeable biomass of biofuel and bio-products. J Chem Technol Biotechnol 2012, 87:11-20.
5. Himmel ME, Ding SY, Johnson DK, Adney WS, Nimlos MR, Brady JW, Foust TD: Biomass recalcitrance: engineering plants and enzymes for biofuels production. Science 2007, 315:804-807.
6. Sindhu R, Binod P, Satyanagalakshmi K, Janu KU, Sajna KV, Kurien N, Sukumaran RK, Pandey A: Formic acid as a potential pretreatment agent for the conversion of sugarcane bagasse to bioethanol. Appl Biochem Biotechnol 2010, 162:2313-2323.
7. Velmurugan R, Muthukumar K: Utilization of sugarcane bagasse for bioethanol production: sono-assisted acid hydrolysis approach. Bioresour Technol 2011, 102:7119-7123.
8. Rezende CA, de Lima MA, Maziero P, de Azevedo ER, Garcia W, Polikarpov I: Chemical and morphological characterization of sugarcane bagasse submitted to a delignification process for enhanced enzymatic digestibility. Biotechnol Biofuels 2011, 4:54.
9. Siqueira G, Milagres AMF, Carvalho W, Koch G, Ferraz A: Topochemical distribution of lignin and hydroxycinnamic acids in sugar-cane cell walls and its correlation with the enzymatic hydrolysis of polysaccharides. Biotechnol Biofuels 2011, 4:7.
10. Taylor CB, Talib MF, McCabe C, Bu L, Adney WS, Himmel ME, Crowley MF, Beckham GT: Computational investigation of glycosylation effects on a family 1 carbohydrate-binding module. J Biol Chem 2012, 287:3147-3155.
11. Scordia D, Cosentino SL, Lee JW, Jeffries TW: Dilute oxalic acid pretreatment for biorefining giant reed (Arundo donax L.). Biomass Bioener 2011, 35:3018-3024.
12. Scordia D, Cosentino SL, Jeffries TW: Second generation bioethanol production from Saccharum spontaneum L. ssp. aegyptiacum (Willd.) Hack. Bioresour Technol 2010, 101:5358-5365.

13. Lee J-W, Houtman CJ, Kim H-Y, Choi I-Y, Jeffries TW: Scale-up study of oxalic acid pretreatment of agricultural lignocellulosic biomass for the production of bio-ethanol. Bioresour Technol 2011, 102:7451-7456.

14. Li X, Cai Z, Horn E, Winandy JE: Effect of oxalic acid pretreatment of wood chips on manufacturing medium-density fiberboard. Holzforschung 2011, 65:737-741.

15. Jeffries TW: Engineering yeasts for xylose metabolism. Curr Opin Biotechnol 2006, 17:320-326.

16. Chandel AK, Singh OV, Chandrasekhar G, Rao LV, Narasu ML: Key-drivers influencing the commercialization of ethanol based biorefineries. J Comm Biotechnol 2010, 16:239-257.

17. Chandel AK, Kapoor RK, Singh AK, Kuhad RC: Detoxification of sugarcane bagasse hydrolysate improves ethanol production by Candida shehatae NCIM 3501. Bioresour Technol 2007, 98:1947-1950.

18. Li Y, Park JY, Shiroma R, Ike M, Tokuyasu K: Improved ethanol and reduced xylitol production from glucose and xylose mixtures by the mutant strain of Candida shehatae ATCC 22984. Appl Biochem Biotechnol 2012, 166:1781-1790.

19. Martín C, Galve M, Wahlbom F, Hagerdal BH, Jonsson LJ: Ethanol production from enzymatic hydrolysates of sugarcane bagasse using recombinant xylose–utilising Saccharomyces cerevisiae. Enzyme Microb Technol 2002, 31:274-282.

20. Santos VTO, Esteves PJ, Milagres AMF, Carvalho W: Characterization of commercial cellulases and their use in the saccharification of a sugarcane bagasse sample pretreated with dilute sulfuric acid. J Ind Microbiol Biotechnol 2011, 38:1089-1098.

21. Esteghlalian A, Hashimoto AG, Fenske JJ, Penner MH: Modeling and optimization of the dilute-sulfuric-acid pretreatment of corn stover, poplar and switch grass. Bioresour Technol 1997, 59:129-136.

22. Rodrigues RCLB, Kenealy WR, Jeffries TW: Xylitol production from DEO hydrolysate of corn stover by Pichia stipitis YS-30. J Ind Microbiol Biotechnol 2011, 38:1649-1655.

23. Mtui GYS: Oxalic acid pretreatment, fungal enzymatic saccharification and fermentation of maize residues to ethanol. Afr J Biotechnol 2011, 11:843-851.

24. Chandel AK, Singh OV, Narasu ML, Rao LV: Bioconversion of Saccharum spontaneum (wild sugarcane) hemicellulosic hydrolysate into ethanol by mono and co-cultures of Pichia stipitis NCIM3498 and thermotolerant Saccharomyces cerevisiae VS3. New Biotechnol 2011, 28:593-599.

25. Chandel AK, Singh OV, Chandrasekhar G, Rao LV, Narasu ML: Bioconversion of novel substrate, Saccharum spontaneum, a weedy material into ethanol by Pichia stipitis NCIM3498. Bioresour Technol 2011, 102:1709-1714.

26. Saha BC, Iten LB, Cotta MA, Wu YV: Dilute acid pretreatment, enzymatic saccharification and fermentation of wheat straw to ethanol. Proc Biochem 2005, 40:3693-3700.

27. Zheng Y, Pan Z, Zhang R, Wang D: Enzymatic saccharification of dilute acid pretreated saline crops for fermentable sugar production. Appl Ener 2009, 86:2459-2467.

28. Chundawat SPS, Donohoe BS, Sousa LD, Elder T, Agarwal UP, Lu FC, Ralph J, Himmel ME, Balan V, Dale BE: Multi-scale visualization and characterization of

lignocellulosic plant cell wall deconstruction during thermochemical pretreatment. Ener Environ Sci 2011, 4:973-984.

29. Kristensen JB, Thygesen LG, Felby C, Jørgensen H, Elder T: Cell-wall structural changes in wheat straw pretreated for bioethanol production. Biotechnol Biofuels 2008, 1/5:1-9.

30. Selig MJ, Viamajala S, Decker SR, Tucker MP, Himmel ME, Vinzant TB: Deposition of lignin droplets produced during dilute acid pretreatment of maize stems retards enzymatic hydrolysis of cellulose. Biotechnol Prog 2007, 23:1333-1339.

31. Igarashi K, Uchihashi T, Koivula A, Wada M, Kimura S, Okamoto T, Penttilä M, Ando T, Samejima M: Traffic jams reduce hydrolytic efficiency of cellulase on cellulose surface.

32. Science 2011, 333:1279-1282. PubMed Abstract | Publisher Full Text OpenURL

33. Wiley JH, Atalla RH: Band assignments in the Raman spectra of celluloses. Carbohyd Res 1987, 160:113-129.

34. Agarwal UP, Weinstock IA, Atalla RH: FT Raman spectroscopy for direct measurement of lignin concentrations in kraft pulps. Tappi J 2003, 2:22.

35. Ooi BG, Rambo AL, Hurtado MA: Overcoming the recalcitrance for the conversion of Kenaf pulp to glucose via microwave-assisted pre-treatment processes. Int J Mol Sci 2011, 12:1451-1463.

36. Colom X, Carrillo F, Nogués F, Garriga P: Structural analysis of photodegraded wood by means of FTIR spectroscopy. Polym Degrad Stab 2003, 80:543-549.

37. Oh SY, Yoo D, Shin Y, Kim HC, Kim HY, Chung YS, Park WH, Youk JH: Crystalline structure analysis of cellulose treated with sodium hydroxide and carbon dioxide by means of X-ray diffraction and FTIR spectroscopy. Carbohyd Res 2005, 340:2376-2391.

38. Pandey KK, Pitman AJ: FTIR studies of the changes in wood chemistry following decay by brown-rot and white-rot fungi. Int Biodet Biodeg 2003, 52:151-160.

39. Pandey KK: A study of chemical structure of soft and hardwood and wood polymers by FTIR spectroscopy. J Appl Polym Sci 1999, 12:1969-1975.

40. Kondo T: The assignment of IR absorption bands due to free hydroxyl groups in cellulose. Cellulose 1997, 4:281-292.

41. Cao Y, Huimin T: Structural characterization of cellulose with enzymatic treatment. J Mol Str 2004, 705:189-193.

42. Sun Y, Lin L, Deng HB, Li JZ, He BH, Sun RC: Structural changes in Bamboo cellulose in formic acid. BioRes 2008, 3:297-315.

43. Krongtaew C, Meesner K, Ters T, Fackler K: Qualitative NIR and pretreatment. BioRes 2010, 5:2063-2080.

44. Belini UL, Hein PRG, Filho MT, Rodrigues JC, Chaix G: NIR for bagasse content of MDF. BioRes 2011, 6:1816-1829.

45. Inagaki T, Siesler HW, Mitsui K, Tsuchikawa S: Crystal structure of cellulose in wood. Biomacromol 2010, 11:2300-2305.

46. Sreenath HK, Koegel RG, Moldes AB, Jeffries TW, Straub RJ: Ethanol production from alfalfa fiber fractions by saccharification and fermentation. Proc Biochem 2001, 36:1199-1204.

47. Abbi M, Kuhad RC, Singh A: Bioconversion of pentose sugars to ethanol by free and immobilized cells of Candida shehatae NCL-3501: fermentation behaviour. Proc Biochem 1996, 31:555-560.

48. Sanchez G, Pilcher L, Roslander C, Modig T, Galbe M, Liden G: Dilute-acid hydrolysis for fermentation of the Bolivian straw material Paja brava. Bioresour Technol 2004, 93:249-256.

49. Sreenath HK, Jeffries TW: Production of ethanol from wood hydrolyzates by yeasts. Bioresour Technol 2000, 72:253-260.

50. Miller GL: Use of dinitrosalicylic acid reagent for determination of reducing sugar. Anal Chem 1959, 31:426-428.

51. Cadete RM, Melo MA, Dussán KJ, Rodrigues RC, Silva SS, Zilli JE, Vital MJ, Gomes FC, Lachance MA, Rosa CA: Diversity and physiological characterization of D-xylose-fermenting yeasts isolated from the Brazilian amazonian forest. PLoS One 2012, 7:e43135.

52. Parekh SR, Yu S, Wayman M: Adaptation of Candida shehatae and Pichia stipitis to wood hydrolysates for increased ethanol production. : Aspen Bibl; 1986:Paper 3623.

53. Gouveia ER, Nascimento RT, Maior AMS, Rocha GJM: Validação de metodologia para a caracterização química de bagaço de cana-de-açúcar. Quim Nova 2009, 32:1500-1503.

54. Scalbert A, Monties B, Janin G: Tannins in wood: comparison of different estimation methods. J Agric Food Chem 1989, 37:1324-1329.

55. Segal L, Creely JJ, Martin AE Jr, Conrad CM: An empirical method for estimating the degree of crystallinity of native cellulose using the X-ray diffractometer. Tex Res J 1962, 29:786-794.

CHAPTER 7

COMBINED BIOLOGICAL AND CHEMICAL PRETREATMENT METHOD FOR LIGNOCELLULOSIC ETHANOL PRODUCTION FROM ENERGY CANE

V. SRI HARJATI SUHARDI, BIJETA PRASAI, DAVID SAMAHA, AND RAJ BOOPATHY

7.1 INTRODUCTION

Environmental impacts of petroleum exploration, as well as the increasing price of oil and gas, necessitate an alternative energy solution [1]. Lignocellulosic biomass is a promising alternative source of energy because of a national abundance of renewable and sustainable feedstocks [2,3]. Biofuels produced from lignocellulosic biomass will enhance national security and stimulate the economy, create jobs, and reduce global climate change. Biomass refers to grasses, agricultural and woody residues, and wastes that can be converted to fuels, chemicals, and electricity [2]. Sugarcane is one of the most efficient crops in converting sunlight energy to chemical energy for fuel [4]. Brazil uses sugarcane as an important energy crop, converting the raw sugar into ethanol. Sugarcane is Louisiana's leading agricultural row crop, worth over $600 million in 2008 [5].

The introduction of energy cane varieties to Louisiana sugarcane farmers could be the forefront of a competitive edge of the sugarcane industry. The new energy cane varieties are a promising development for cellulosic ethanol production. In 2007, three energy cane varieties were released by the United States Department of Agriculture (USDA), namely, L 79-1002, HoCP 91-552, and Ho 00-961 [4]. A current commercial variety of sugarcane is HoCP 96-540. Energy cane can withstand freeze and it is a 12 month crop compared to commercial sugarcane which is a nine month crop and it has to be harvested before freeze set in every year. Energy cane also can grow in poor agricultural soil and it does not compete with food crop [5]. The energy cane variety L 79-1002 is more suitable for ethanol production compared to other varieties of energy cane because it contains significantly higher cellulose and hemicellulose content than any other sugar cane [6].

Lignocellulosic biomass consists of a network of cellulose and hemicellulose bound by lignin. The process of converting biomass to ethanol involves pretreatment to remove lignin and free sugars followed by enzymatic saccharification and fermentation. The lignin sheath as well as the crystallinity of cellulose presents major challenges to these pre-treatment techniques. However, alkaline [7-10] and dilute acid solutions [8-12] can effectively remove lignin and reduce cellulose crystallinity. Determining the optimal pre-treatment for energy cane varieties is necessary to develop efficient fermentation for ethanol production.

The release of cellulose and hemicellulose allows for post-treatment enzymatic saccharification of these carbohydrates to simple sugars for fermentation. The more effective the pretreatment is at loosening the crystallinity of lignocellulosic biomass the more carbohydrates are available for enzymatic saccharification, thereby increasing Renewable Bioresources Combined biological and chemical pretreatment method for lignocellulosic ethanol production from energy cane ethanol yield from fermentation [13,14]. In this study the biomass used was sugarcane leaf from the energy cane. Every year after sugarcane is harvested, farmers typically reduce residue by open air burning. This is a cost-effective way to remove the fibrous content that would otherwise significantly reduce milling efficiency and decrease profits, as well as to clear residue from the field that hinders farming [9]. The open air burning practice not only affects the quality of air but also the quality of life to those who live in the area. One alternative to open air burning is the production of ethanol from sugarcane residue. Ethanol is a clean burning, renewable resource that can be produced from cellulosic biomass. The main purpose of this study was to find an economical and best pretreatment method for a particular energy cane variety L 79-1002 for lignocellulosic ethanol production.

7.2 MATERIALS AND METHODS

7.2.1 MATERIALS

Leaves of energy cane varieties L 79-1002 was collected in May and June of 2010 from the United States Department of Agriculture (USDA) sugarcane research unit in Houma, LA. Leaf tops were cut in three to five centimeter pieces and stored in muck buckets in the laboratory. A recombinant *Escherichia coli* FBR 5 was kindly provided by Dr. Mike Cotta of National Center for Agricultural Utilization Research of USDA, Peoria, IL, USA. This recombinant *E. coli* is known to ferment both glucose and xylosic sugars from cellulose and hemicellulose of wheat hydrolysate [15]. Brown rot and white rot fungi, namely, *Cerioporiopsis pannocinta* (ATCC 9409) and *Phanerochaete chrysosporium* (ATCC 32629) were obtained from the American Type Culture Collection (ATCC; Manassas, VA). All chemicals used in the study were of reagent grade. *E.coli* was maintained in LB broth medium and the fungi were maintained in potato dextrose agar (PDA) medium. Cellulase, β-glucanase, and endo-1,4-β-xylanase enzymes were from Sigma chemicals, St. Louis, MO.

7.2.2 PRE-TREATMENT

Dilute acid pretreatments at moderate temperatures free hemicellulose and cellulose [11] and disrupt lignin, thereby releasing cellulose for enzymatic reactions [16]. In this study 0, 1, 2, 3, and 4% H_2SO_4 solutions were used for pretreatment of lignocellulosic biomass.

Energy cane L 79-1002 was cut into 2-5 cm pieces and dried in an oven at 100°C for six hours to remove any moisture. Ten grams of the dry energy cane were placed into each labeled anaerobic bottle. Different concentrations of H_2SO_4 solution were added so that the energy cane was submerged (150 mL). The volumetric ratio of biomass to dilute acid was 1:15 (10 g biomass and 150 mL dilute acid). All acid treatments were done in triplicate as well as the control, which used DI water. Each sample was soaked for 24 hours in respective concentrations of H_2SO_4 and then autoclaved at 121°C for 20 minutes at 15 psi. The H_2SO_4 solution was removed, and each sample was triple rinsed with DI water for a total of three hours (one rinse per hour).

The fungal pretreatment was performed in solid state fermentation (SSF) using a sterile Ziploc bag filled with 10 gram of dry energy cane cut into 2-5 cm pieces as described in detail by Lyn et al., [17]. Fungal treatment includes individual fungus alone and combination of both fungi with a total of three treatments and each treatment had triplicates. Pregrown fungi were inoculated into the Ziploc bags as an agar plug grown on PDA for three days with 100% coverage of mycelium on the agar surface. A 5% (W/W) agar plug was used as inoculum. The bags were maintained with 70% moisture (analyzed using a moisture analyzer, Fisher Scientific, St. Louis, MO) and incubated for 10 days at room temperature (20-22°C) to simulate the biomass storage conditions prior to processing for biofuel in a large-scale production unit.

7.2.3 COMBINATION OF FUNGAL AND ACID PRE-TREATMENT

An experiment was conducted with a fungal pretreated biomass with both fungi as described in pre-treatment section. The fungal pretreated biomass was subjected to dilute acid pretreatment with low concentrations of acids, namely, 0.25, 0.5, 1, 1.5, and 2% sulfuric acid as described above. These various combined pretreated biomasses underwent enzymatic saccharification and fermentation as described in enzymatic saccharification and fermentation sections.

7.2.4 ENZYMATIC SACCHARIFICATION AND FERMENTATION

The pre-treated biomass from dilute acid, fungal, and combined pre-treatments were subjected to separate hydrolysis and fermentation (SHF). Pretreated samples underwent SSF with enzymatic saccahrification for 18 hours at 30°C with the addition of cellulase enzymes (Sigma C9748), β-glucanase (Sigma G4423), and hemicellulose enzyme endo-1,4-β-xylanase (Sigma X2629) at 10% protein of enzyme dosing of each enzyme as described by Shields and Boopathy [6]. The enzyme activity was equal to three filter paper units (FPU) per 1% protein of the enzyme. After 18 hours of enzyme reaction, a 5% recombinant E.coli FBR 5 pregrown in LB medium with the optical density of 1.2 at 600nm was introduced into individual treated bottles to start the fermentation. The fermentation medium was basic mineral salt medium with the volume of 150 mL in 250 mL anaerobic bottle as described by Shields and Boopathy [6]. The fermentation was carried out anaerobically without shaking in the anaerobic bottles. The initial oxygen in the headspace was rapidly consumed within six hours of incubation. The initial pH of the medium was 6.0 and the fermentation temperature was 30°C. Samples were periodically drawn for ethanol analysis. The fermentation lasted for six days.

7.2.5 ETHANOL ANALYSIS

All fermentation samples were analyzed for ethanol production using high performance liquid chromatography (HPLC) as described by Dawson and Boopathy [9] and Shields and Boopathy [6]. A Varian Pro Star Autosampler Model 410 liquid chromatograph equipped with two solvent pumps and Infinity UV and diode array detector with a data module, and a model 320 system controller were used. The mobile phase was 0.0025 N H_2SO_4. Aliquots of 10 μL were injected into an organic acid column (Varian organic acid column, Cat #SN 035061) at 22°C. The flow rate of the mobile phase was 0.6 mL/min, and the analysis was done under isocratic mode. An ethanol standard was used for quantification of ethanol in the sample.

7.2.6 STATISTICAL ANALYSIS

Analysis of variance (ANOVA), followed by a Tukey post-hoc range test (p < 0.05) as described by Neter et al., [18] was used to analyze ethanol production.

7.3 RESULTS AND DISCUSSION

Two sets of pretreatment were studied in detail: one with different concentrations of dilute sulfuric acid and another with two kinds of fungi under individual and combined fungal treatment conditions. The results are presented in Table 1. The dilute acid pretreatment showed an increase in ethanol yield as the acid concentration increased from 1% to 3%. Further increase in acid concentration to 4% showed a decrease in ethanol yield compared to 3% acid treatment. This may be due to the presence of inhibitory compounds such as furfural and 5-hydroxy methylfurfural. Several reports indicate the presence of these inhibitory compounds at higher acid concentrations [6,9]. The maximum ethanol concentration of 3021 mg/L was observed in 3% sulfuric acid treatment, which was significant compared to other concentrations used in the study and also other studies reported in the literature including Dawson and Boopathy [9,10] and Shields and Boopathy [6].

Table 1 also shows ethanol yield among various fungal pretreatments. The best result was achieved in a combined pretreatment of *Cerioporiopsis* (brown rot fungus) and *Phanerochaete* (white rot fungus). The ethanol yield in this condition was 1345 mg/L, which was significantly less than dilute acid pretreatment. Individual fungal pretreatment produced lower ethanol yield. In natural systems, fungi, especially the brown rot and white rot fungi, are known to decompose fallen leaves from trees and other plants to humic and water soluble compounds [17]. These fungi produce various enzymes such as lignin peroxidase, phenol oxidase, manganese peroxidase, and laccase [19-21]. These enzymes can be produced both under submerged fermentation (SmF) and solid-state fermentation (SSF) [22]. In this study, the SSF pretreatment showed effective release of cellulose and

TABLE 1: Effect of Various Pretreatments on Ethanol Production after Six Days of Fermentation.

Pretreatment	Ethanol Production (mg/L)
Dilute acid pretreatment: Control (no acid treatment)	15 + 0.6
1% sulfuric acid	877 + 21.5A
2% sulfuric acid	1725 + 32.1A
3% sulfuric acid	3021 + 27.8AB
4% sulfuric acid	2876 + 39.5AB
Fungal pretreatment: Control (no fungus)	2 + 0.7
Cerioporiopsis alone	687 + 7.5C
Phanerochaete alone	766 + 5.6C
Cerioporiopsis+Phanerochaete	1345 + 14.9CD

Results are average of triplicates in each treatment with S.D. Data with similar letters are not significantly different from each other under two different sets of pretreatment conditions.

The pretreated energy cane was subjected to enzymatic saccharification and fermentation with recombinant E.coli FBR 5 as detailed in methods section.

hemicellulose, which resulted in significantly higher ethanol production in the fungal pretreated energy cane compared to control.

Based on the results obtained from two different pretreatments, further experiments were carried out to combine the dilute acid pretreatment with fungal pretreatment in order to reduce the use of acid, which will be a big cost factor in large scale biofuel production systems. Energy cane was subjected to a pretreatment condition with *Cerioporiopsis* and *Phanerochaete* together, which yielded higher ethanol yield among various fungal pretreatments (Table 1) as detailed in method section. Following ten days of fungal pretreatment, the energy cane was pretreated with various low concentrations of sulfuric acid (0, 0.25, 0.5, 1, 1.5, and 2%). The pretreated biomass was enzymatically saccharified and subjected to fermentation using recombinant E.coli FBR 5. The results from this study are given in Table 2. The energy cane with 0% sulfuric acid after 10 days of fungal

TABLE 2: Effect of Fungal Pretreatment on Dilute Acid Pretreatments in Ethanol Production after Six Days of Fermentation.

Treatment	Ethanol Production (mg/L)
0% sulfuric acid	1266 ± 11.5A
0.25% sulfuric acid	1325 ± 22.7A
0.5% sulfuric acid	1971 ± 29.5A
1% sulfuric acid	2876 ± 39.2AB
1.5% sulfuric acid	2956 ± 41.2AB
2% sulfuric acid	3055 + 25.3AB

Results are average of triplicates in each treatment with S.D. Data with similar letters are not significantly different from each other.

Energy cane was pretreated with Cerioporiopsis and Phanerochaete for 10 days followed by various dilute acid treatments before the hydrolysate was subjected to enzymatic saccharification and fermentation with recombinant *E.coli* FBR 5 as detailed in methods section.

treatment produced ethanol concentration of 1266 mg/L compared to 2% sulfuric acid treatment of fungal pretreated biomass, which produced 3055 mg/L of ethanol (p value = 0.01). However, the lower dilution of 1 and 1.5% produced equally good amount of ethanol, namely, 2876 and 2956 mg/L, respectively. Statistical analysis showed no significant difference among 1, 1.5, and 2% dilute acid treatment of fungal pretreated energy cane with a p value of 0.32.

Pretreatment of lignocellulosic biomass is a costly step [23], but is essential for high ethanol yields on a commercial level. Efficient pretreatment can affect downstream process costs by reducing the use of enzymes or fermentation time [23]. In our previous studies, we reported acid pretreatment was better than alkaline pretreatment in removing lignin from commercial sugarcane residues such as leaf and bagasse [6,9,10]. In the current study, acid pretreatment with fungal treatment of biomass was chosen as the best pretreatment to release cellulose and hemicelluose from lignin in the leaf biomass of energy cane L79-1002. The results from the two experiments showed that the combination fungal pretreatment with very dilute sulfuric acid (1%) of energy cane produced 2876 mg/L of ethanol and the ethanol production was 3021 mg/L in the treatment that received 3% sulfuric acid without fungal treatment (Tables 1 and 2). The ethanol production in these two treatments are almost similar. Thus combining the fungal treatment with dilute acid treatment could save a significant volume of acid that is needed for pretreatment of energy cane for ethanol production. This difference of 2% acid volume is significant and could be practical in the large scale bioprocessing of lignocelluosic materials for biofuel production as the biomass can be treated with fungi during storage period prior to biomass processing. Further research is needed in scaling up the process, which will help us to do economical analysis for biofuel industry.

7.4 CONCLUSIONS

This study shows that dilute acid pretreatment released cellulose and hemicellulose, which are available for enzymatic saccharification and fermentation. The best dilute acid pretreatment was 3% sulfuric acid.

The use of fungal pretreatment enhanced ethanol production. Brown rot and white rot fungi produced almost similar ethanol yield. The combined treatment of brown rot and white rot fungi together produced significantly higher ethanol yield compared to control, however, produced less ethanol compared to 3% dilute sulfuric acid pretreatment.

The combination of fungal pretreatment with lower dilute acid pretreatment produced the best result of this study. A 10 day fungal pretreated energy cane with both brown rot and white rot fungi together treated with 1% sulfuric acid showed ethanol production of 2876 mg/L, which is comparable to ethanol production in 3% dilute acid treatment without fungal pretreatment and thus combining the fungal pretreatment with acid pretreatment makes practical sense.

REFERENCES

1.   Cobill RM: Development of energy canes for an expanding biofuels industry. Sugar J 70: 6.

2. U.S. DOE: Breaking the biological barriers to cellulosic ethanol: a joint research agenda, DOE/SC/EE-0095, U.S. Department of Energy Office of Science and Office of Energy Efficiency and Renewable Energy. 2006.

3. U.S. DOE: Biomass: multi-year program plan. U.S. Department of Energy Office of Energy Efficiency and Renewable Energy. 2009.

4. Tew T and Cobill R: Genetic improvement of sugarcane (Saccharum spp.) as an energy crop. In: Vermerris, W. (Ed.), Genetic Improvement of Bioenergy Crops. Springer Science + Business Media, LLC, New York, NY, 2008. pp. 249-272.

5. Salassi M, Deliberto M and Legendre B: Economic importance of Louisiana sugarcane production in 2008. LSU AG Center. 2009.

6. Shields S and Boopathy R: Ethanol production from lignocellulosic biomass of energy cane. Internationat Biodet & Biodeg 2011, 65:142-146.

7. Gould JM: Alkaline peroxide delignification of agricultural residues to enhance enzymatic saccharification. Biotechnol Bioeng 1984, 26:46-52.

8. Gould JM: Studies on the mechanism of alkaline peroxide delignification of agricultural residues. Biotechnol Bioeng 1985, 27:225-31.

9. Dawson L and Boopathy R: Use of post-harvest sugarcane residue for ethanol production. Bioresour Technol 2007, 98:1695-9.

10. Dawson L and Boopathy R: Cellulosic ethanol production from sugarcane bagasse without enzymatic saccharification. BioResources 2008, 3: 452-460.

11. Knappert DR, Grethlein HE and Converse AO: Partial acid hydrolysis of poplar wood as a pretreatment for enzymatic hydrolysis. Biotech & Bioeng 1981, 11: 67–77.

12. Grohmann K, Torget R and Himmel M: Dilute acid pretreatment of biomass at high solids concentration. Biotech & Bioeng Symp 1986, 17: 135-151.

13. Hari Krishna S and Chowdary GV: Optimization of simultaneous saccharification and fermentation for the production of ethanol from lignocellulosic biomass. J Agric Food Chem 2000, 48:1971-6.

14. Chapple C, Ladisch M and Meilan R: Loosening lignin's grip on biofuel production. Nat Biotechnol 2007, 25:746-8.

15. Saha BC and Cotta MA: Continuous ethanol production from wheat straw hydrolysate by recombinant ethanologenic Escherichia coli strain FBR5. Appl Microbiol Biotechnol 2011, 90:477-87.

16. Yang B and Wyman CE: Effect of xylan and lignin removal by batch and flowthrough pretreatment on the enzymatic digestibility of corn stover cellulose. Biotechnol Bioeng 2004, 86:88-95.

17. Lyn M, Boopathy R, Boykin D, Weaver MA, Viator R and Johnson R: Sugarcane residue decomposition by white rot and brown rot microorganisms. Sugarcane Internat J 2010, 28:37-42.

18. Neter J, Wasserman W and Kutner MH: Applied linear statistical models: regression, analysis of variance, and experimental designs. 3rd ed. 1990, IRWIN. Burr Ridge, Illinois.

19. Kuhad RC, Singh A and Eriksson KE: Microorganisms and enzymes involved in the degradation of plant fiber cell walls. Adv Biochem Eng Biotechnol 1997, 57:45-125.

20. Leonowicz A, Matuszewska A, Luterek J, Ziegenhagen D, Wojtas-Wasilewska M, Cho NS, Hofrichter M and Rogalski J: Biodegradation of lignin by white rot fungi. Fungal Genet Biol 1999, 27:175-85.

21. Howard RL, Abotsi E, Rensburg JEL and Howard S: Lignocellulosic biotechnology: issues of bioconversion and enzyme production. African Journal of Biotechnology 2003, 2:602-619.

22. Osma JF, Herrera JLT and Couto SR: Banana skin; A novel waste for laccase production by Trametes pubescens under solid state conditions application to synthetic dye decoloration. Dyes and Pigments 2007, 75:32-37.

23. Lynd LR, Elander RT and Wyman CE: Likely features and costs of mature biomass ethanol technology. Appl Biochem & Biotechnol 1996, 57/58: 741-761.

CHAPTER 8

A NOVEL PROMISING *TRICHODERMA HARZIANUM* STRAIN FOR THE PRODUCTION OF A CELLULOLYTIC COMPLEX USING SUGARCANE BAGASSE *IN NATURA*

BRUNO BENOLIEL, FERNANDO ARARIPE GON3ALVES TORRES, AND LIDIA MARIA PEPE DE MORAES

8.1 INTRODUCTION

Lignocellulosic residues derived from different agro-industrial activities represent a massive source of raw material for the production of fuels, chemical feedstock, foods and livestock feeds (Kumar et al. 2008). Brazil is a major producer of renewable feedstock including sugarcane which is essentially used for sugar and fuel ethanol production. In 2010, sugarcane production reached ~717.5 million tons (FAOSTAT 2012). A significant fraction of this biomass goes to industries for steam and electricity generation. The remaining fraction represents the ideal feedstock for the genera-

A Novel Promising Trichoderma harzianum Strain for the Production of a Cellulolytic Complex Using Sugarcane Bagasse in Natura. © *2013 Benoliel et al.* Springerplus, *2013; 2: 656.; licensee Springer.*

tion of high-value commodities as second-generation ethanol (Canilha et al. 2012).

The use of lignocellulosic biomass for the production of second generation ethanol requires a pretreatment for the liberation of carbohydrate polymers. A number of different strategies have been envisioned to convert the polysaccharides into fermentable sugars. One of them is accomplished by weak acid (chloridric or sulfuric acid) treatment (Betancur and Pereira Jr. 2010) to hydrolyze the hemicellulose fraction. The resulting solid fraction is then depolymerized by a chemical or enzymatic treatment. The later involves the use of different classes of hydrolytic enzymes generally produced by filamentous fungi. The conversion of cellulose to glucose involves the concerted action of three classes of enzymes: endo-β-1,4-glucanases (EC 3.2.1.4), exo-cellobiohydrolases (EC 3.2.1.91), and β-glucosidases (β-D-glucosidic glucohydrolases, EC 3.2.1.21). Hydrolytic enzymes represent a considerable cost in industrial biofuel plants. On-site enzyme production has been proposed as a way of lightening this burden. A fraction of lignocellulosic material partially hydrolyzed is diverted from the process and used as a cheap carbon source for enzyme production. Consequent reductions in enzyme storage time and downstream processing can thus lower the overall costs (Himmel et al. 1999; Tolan 2002).

The capacity of a particular microorganism to grow in lignocellulosic substrates is directly related to the production of a broad spectrum of enzymes that act synergistically to deconstruct the plant cell wall by depolymerization of substrates of different complexities (Andreaus et al. 2008; Kumar et al. 2008; Siqueira et al. 2010; Moreira et al. 2012). An assorted library of cellulolytic microbes should facilitate the development of optimal enzyme cocktails specific for locally available lignocellulosic biomass, such as sugarcane bagasse.

The Cerrado is the main savanna-like region in the Americas covering about 2 million km^2. Although considered an important biodiversity hotspot (Myers et al. 2000) its microbial diversity has not been thoroughly assessed for biotechnological purposes. Our group has previously isolated an xylanase producing yeast from de Brazilian Cerrado (Parachin et al. 2009). The objective of this work was to evaluate the production of cellulases by a set of filamentous fungi isolated from the Brazilian Cerrado aiming their use as a potential source for on-site enzyme production.

8.2 MATERIALS AND METHODS

8.2.1 MICROORGANISMS AND FUNGAL ISOLATION

Trichoderma reesei Rut-C30 (ATCC 56765) (Montenecout and Eveleigh 1977) was used as reference strain. Decaying leaf litter encountered in the Cerrado's soil from different areas around the city of Brasília was used as a source for fungal isolation. Collected samples were immersed in sterile distilled water and after vigorous agitation the suspension was subjected to serial dilutions and plated in a medium similar to the basic nutrient medium of Mandels and Weber (1969), with the exception that urea was omitted, a double amount of $(NH_4)_2SO_4$ was included, and the peptone content was elevated by 20% (Szijártó et al. 2004), supplied with 1% carboxymethyl cellulose. Plates with standard inoculum of 103 spores were incubated for 5 days at 30°C following Congo red staining (Ruegger and Tauk-Tornisielo 2004). Positive cellulolytic colonies were picked up and subcultured on potato dextrose agar (PDA - 0.4% potato, 2% dextrose, 0.5% peptone, 2% agar) slants and grown at 30°C for 10 days.

Submerged fermentations were performed in 500 mL Erlenmeyer flasks. Total of 106 spores were inoculated into 100 mL of modified basic nutrient Mandels and Weber medium (Szijártó et al. 2004) supplied with 10 g.L-1 of different carbon sources. The cultures were incubated on a rotary shaker with an agitation rate of 200 rpm at 30°C for up to 120 hours. Throughout the cultivation, aliquots were withdrawn, centrifuged at 20,000 g for 5 min for cell and residual substrate analysis. Supernatants were stored at -20°C until the enzyme assays were carried out. Three biological replicates were performed for each condition.

8.2.2 ENZYME AND PROTEIN ASSAYS

Total cellulase, endoglucanase, and exoglucanase activities were determined using Whatman no. 1 filter paper (FPA), carboxymethyl cellulose (CMC, low viscosity), and microcrystalline cellulose (SIG) as substrates, respectively, according to standard conditions described by Ghose (1987).

Reducing sugars, expressed as glucose liberated during reactions on FPA, CMC and SIG were quantified by the DNS method (Miller 1959). Endoxylanase activities were determined using xylan from oat spelts (XYL) as substrate according to Bailey et al. (1992). For all reactions one enzyme unit (U) was defined as the amount of biocatalyst that releases 1 μmol of the correspondent monosaccharide (xylose for xylanase and glucose for the other groups of enzymes) per minute under the assay conditions (30 min incubation at 50°C with 50 mM acetate buffer pH 5.0). β-Glucosidase activity was assayed in a 100 μL reaction mixture containing 3 mM p-nitophenyl-β-D-glucopyranoside (pNPG; Sigma, St. Louis, USA), 50 mM acetate buffer (pH 5.0), and an appropriate dilution of enzyme preparation. After 10 min incubation at 50°C, the reaction was stopped by adding 200 μL of 1 M Na2CO3, and p-nitrophenol (pNP) release was monitored at A405nm. Enzyme unit was defined as the amount of biocatalyst that releases 1 μmol p-nitrophenol per minute under the assay conditions. Total extracellular protein content was measured using the Bio-Rad protein reagent according to the Bradford method (Bradford 1976) using bovine serum albumin (Sigma) as standard. All analyses were done in triplicate in a temperature-controlled incubator.

8.2.3 SUGARCANE BAGASSE HYDROLYSIS

In natura sugarcane bagasse (SCB), used as substrate in the hydrolysis assays, was provided by Costa Pinto Mill (Piracicaba, SP, Brazil). This biomass was also subjected to pretreatments to generate partially delignified cellulignin (PDC). Acid and alkali pretreatments were carried out to increase the cellulose content in the materials by removing the hemicellulose fraction and partially removing the lignin fraction, respectively. The acid pretreatment consisted of incubating the solid material with a 3% (v/v) sulfuric acid solution (solid:liquid ratio of 1:4), while the alkali pretreatment was performed by incubating the material with a NaOH 4% (w/v) solution (solid:liquid ratio of 1:20). Both pretreatments were carried out at 121°C (1 atm) for 20 min.

Enzymatic hydrolysis was carried out using 1.0% partially delignified cellulignin (PDC) or *in natura* sugarcane bagasse (SCB) using a

SCB-induced (72 h) culture supernatant. Suspensions were incubated at 50°C in 50 mM sodium citrate buffer (pH 5.0) for 18 h with regular sampling. Glucose concentration was determined usinga kit based on the glucose oxidase assay (Katal®) and total reducing sugar content was determined according to the method described by Miller (1959). The hydrolysis yield was determined using equation described by Maeda et al. (2011) from the carbohydrate contents of SCB and PDC previously determined by Castro et al. (2010a).

8.2.4 GEL ELECTROPHORESIS AND ZYMOGRAM ANALYSIS

SDS-PAGE analysis was performed to detect extracellular proteins. For visualization of carboxymethyl-cellulase and xylanase activities on gels a zymogram was performed according to Sun et al. (2008) with modifications. Briefly, proteins were separated on a 10% SDS-PAGE gel with either 0.15% CMC (low viscosity sodium salt) or 0.15% XYL (xylan from birch wood). The gel was washed twice in a solution of 0.5 M sodium acetate and 25% isopropanol at room temperature to remove SDS. Proteins were renatured in 50 mM acetate buffer (pH 5.0) containing 5 mM β-mercaptoethanol by stirring the gel overnight at 4°C. The gel was then incubated in 50 mM acetate buffer (pH 5.0) at room temperature for 2 h followed by incubation at 50°C for another 2 h. The gel was stained in 0.2% Congo red for 1 h and distained with 1 M NaCl.

8.2.5 MOLECULAR IDENTIFICATION

Standard protocols were followed for DNA manipulations (Sambrook and Russel 2001). Total fungal DNA was obtained from a mycelium grown on PDA as described by Roeder and Broda (1987). Ribosomal internal transcribed spacer (ITS) region was amplified from genomic DNA by PCR using primers ITS1 (5'-GCGGATCCGTAGGTGAACCTGCGG) and ITS4 (5'-GCGGATCCTCCGCTTATTGATATGC) (White et al. 1990). Double-stranded DNA sequencing was performed with the MegaBACE® Dye Terminator kit (GE Healthcare). Computer sequence analysis was carried out using the Phrap and Phred programs (Ewing et al. 1998).

Fungal isolates were identified via ITS sequence analysis using BLASTn search tools (http://www.ncbi.nlm.nih.gov). For taxonomic considerations, the obtained sequences were also used to include related species into phylogenetic trees. Sequence alignment was carried out using CLUSTALW (http://www.ebi.ac.uk/Tools/msa/clustalo). Phylogenetic analysis was performed using MEGA v 4.0 software (Tamura et al. 2007). Bootstrap resampling analysis for 1000 replicates was performed to estimate the confidence of results. The DNA BarCode method for *Trichoderma* identification was carried out using the TrichOKEY v. 2 program (Druzhinina et al. 2005).

8.3 RESULTS

In a screening for cellulolytic activity a total of 13 isolates (L01-L13) from decaying leaf litter were selected after growth on cellulose as sole carbon source. Sequence analysis of PCR-amplified fungal ITS revealed a wide diversity of fungal taxa representing five genera: *Trichoderma* (5), *Penicillium* (4), *Aspergillus* (2), *Pestalotiopsis* (1) and *Curvularia* (1). All thirteen isolates demonstrated the ability to growth on cellulose as sole carbon source, but only four (L04, L08, L10, L11) showed a significant CMC hydrolysis halo when submitted to a rapid screening for cellulolytic activity on Congo red plate assay. The L04 isolate presented the fastest growth occupying all 9 cm diameter plate area in about 3 days while L10 took 4 days. Isolates L08 and L11 presented slower growth on CMC medium although the best colony/halo ratios, 0.72 and 0.63, respectively. No significant growth and/or activity hydrolysis halo on cellulosic substrate were observed in the other strain plates, therefore L04, L08, L10 and L11 isolates were selected for further studies.

To evaluate the cellulolytic productivity of the four selected isolates, submerged fermentation using CMC as sole carbon source was carried out for up to 120 hours. Every 24 hours, samples were withdrawn and assayed for cellulase activity. All analyzed isolates showed similar exoglucanase productivity when CMC was used as carbon source; the highest values observed were all around 1,200 $U.L^{-1}$ in a culture time of 72 hours.

TABLE 1: Maximum values of volumetric productivity observed for cellulases production by Cerrado isolates.

Strain	Substrate	Endoglucanase	Exoglucanase	β-glucosidase
L04	CMC	30.6±1.2 (72)	25.9±2.0 (24)	24.5±1.0 (96)
	SCB	64.2±6.4 (48)	10.2±1.2 (120)	52.0±0.4 (24)
L08	CMC	20.2±1.2 (24)	24.9±1.5 (48)	ND
	SCB	ND	6.3±0.3 (96)	ND
L10	CMC	8.8±0.3 (24)	30.9±1.2 (24)	ND
	SCB	2.9±0.1 (48)	36.8±1.4 (24)	ND
L11	CMC	10.6±0.7 (24)	25.2±1.7 (48)	4.5±0.2 (48)
	SCB	ND	12.2±0.5 (24)	ND

Volumetric productivity (U. L^{-1} .h^{-1}). Values in parentheses correspond to time of fermentation (h) when maximum results were observed. Not detected activity (ND).

L04 excelled at the endoglucanase productivity when compared to the others analyzed fungi, reaching the value of 2,206 U.L^{-1} (72 h). It represents about 6 times the production of L08, L10 and L11 at the same point of the growth curve. The production of β-glucosidase was also analyzed. The maximum activity detected for L04 was 2,938 U.L^{-1} (120 h), although at 96 hours of culture 2,350 U.L^{-1} has been achieved.

To evaluate the cellulolytic production ability of the selected strain in a complex substrate, SCB was used as carbon source in a submerged fermentation. Strain L04 demonstrated to be more efficient than the others analyzed strains when grown on SCB. The L04 maximum activities for endoglucanase, exoglucanase and β-glucosidase when grown on this substrate were 4,022 U.L^{-1} (72 h), 1,228 U.L^{-1} (120 h) and 1,968 U.L^{-1} (48 h), respectively. SCB seemed as a poor cellulase inductor for the other analyzed strains. No activities or very low endoglucanase and β-glucosidase activities were detected in the supernatant from L08, L10 and L11, and the maximum exoglucanase activity detected reached half of the values obtained by L04 on this substrate.

In terms of volumetric productivity, strain L04 reached the maximum values as early as 24 hours for β-glucosidase (52.0 U.L^{-1}.h^{-1}) and 48 hours for endoglucanase (64.2 U.L^{-1}.h^{-1}) when SCB was used as inducer (Table 1). When *T. reesei* Rut C30 was grown on SCB as carbon source, it presented maximum volumetric productivity values for endoglucanase, exoglucanase and β-glucosidase of 38.6 U.L^{-1}.h^{-1} (72 h), 14.2 U.L^{-1}.h^{-1} (24 h) and 29.3 U.L^{-1}.h^{-1} (48 h), respectively. The *T. reesei* Rut C30 maximum detected activities for endoglucanase, exoglucanase and β-glucosidase when grown on this substrate were 3,795 U.L^{-1} (192 h), 567 U.L^{-1} (120 h) and 1,979 U.L^{-1} (192 h), respectively. Strain L04 was identified as *Trichoderma harzianum* (*Hypocrea lixii*) by phylogenetic analysis of its ITS1/2 regions and was selected for further analysis.

The L04 enzymatic profile produced on SCB presented more xylanase, endoglucanase and β-glucosidase activities, even when grown on specific substrates (Table 2). However, partially delignified cellulignin (PDC) showed to be a poor inducer of L04 cellulolytic system (data not shown).

PDC and SCBwere subjected to enzymatic saccharification using the enzymatic cocktail produced by L04 grown on SCB. The enzyme/biomass loading ratio at this assay was equivalent to 5.0 FPU/g substrate. Figure 1 presents the temporal profile of the concentration of reducing sugars and glucose released by enzyme extract. After 18 h of saccharification, the L04 extract was able to release 4.32 g.L^{-1} of total reducing sugars from SCB and 8.16 g.L^{-1} from PDC. Glucose contents were also measured presenting values of 2.28 g.L^{-1} from SCB and 4.48 g.L^{-1} from PDC after 18 h of substrate hydrolysis. The released glucose concentrations were higher for PDC than SCB, however, in terms of glucose yield after 18 h of hydrolysis, calculated from the cellulose contents of SCB and PDC, the values were not different. The hydrolysis yield determined using SCB and PDC were 60.32% and 59.35%, respectively.

To detect cellulolytic and xylanolytic activities in L04 enzymatic cocktail used in the saccharification assay, electrophoretic analysis on SDS-PAGE and zymographic assays was also performed. The L04 protein profile presented two distinct molecular mass cellulolytic activities, ~50 kDa and 20 kDa (Figure 2, lane 3). Xylanolytic activities were also detected as a defined 20 kDa band and multiple activities over 75 kDa (Figure 2, lane 5).

TABLE 2: Maximum values of volumetric productivity observed for production of cellulases by *T. harzianum* L04 strain grown on different carbon sources.

Substrate	Endoglucanase	Exoglucanase	Total cellulase	β-glucosidase	Xylanase
SCB	64.2±6.4 (48)	10.2±1.2 (120)	15.4±0.2 (48)	52.0±0.4 (24)	135.7±1.7 (48)
CMC	30.6±1.2 (72)	25.9±2.0 (24)	3.8±0.3 (24)	24.5±1.0 (96)	44.5±3.1 (48)
SIG	37.2±1.6 (96)	3.1±0.1 (24)	9.8±0.6 (96)	40.4±3.4 (48)	105.8±5.3 (24)
XIL	19.0±0.9 (24)	9.5±0.6 (24)	3.8±0.4 (48)	4.5±0.3 (96)	132.5±6.1 (24)

Volumetric productivity (U. L^{-1} .h^{-1}). Values in parentheses correspond to time of fermentation (h) when maximum results were observed.

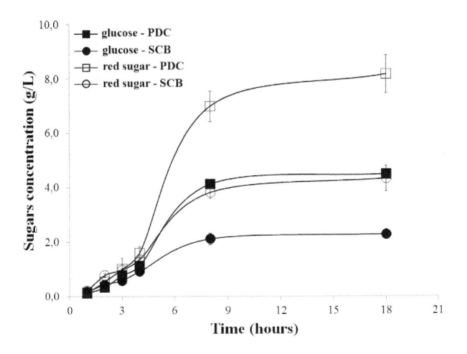

FIGURE 1: Hydrolysis of 10% (w/v) *in natura* sugarcane bagasse (SCB) and partially delignified cellulignin (PDC) using enzyme cocktails produced by *T. harzianum* L04. Concentrations of total reducing sugars and glucose during hydrolysis are indicated.

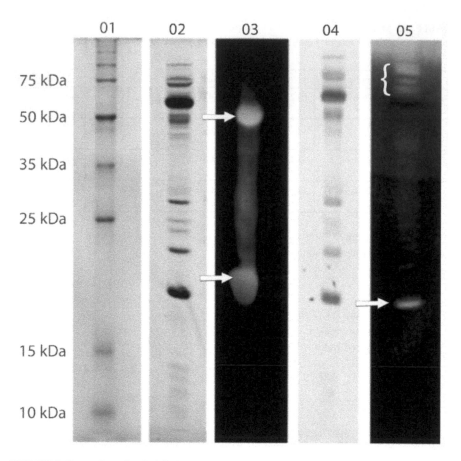

FIGURE 2: Detection of cellulolytic and xylanolytic activities in *T. harzianum* L04 strain grown on SCB. Molecular weights are indicated on lane 01. SDS-PAGE analysis (lanes 2 and 4). For detection of cellulolytic (03) and xylanolytic activities (05), cellulose and xylan were incorporated into the gel, respectively and stained with Congo red. The arrows in 03 indicate cellulolytic activities. The arrow and bracket in 05 the xylanolytic activities.

8.4 DISCUSSION

In nature, the breakdown of plant materials is done primarily by fungi. In its natural environment saprophytic fungi can colonize the leaf litter and woody debris in humus or associated with plant matter in the soil. In order to grow on theses complexes substrates the ability to produce a broad spectrum of protein and polysaccharide hydrolyzing enzymes is required. For this reason, samples were collected from the Brazilian Cerrado environment and the fungi biodiversity analysed .

The wide diversity of fungal taxa found on Cerrado samples and the predominance of *Trichoderma* and *Penicillium* genera were expected, since the technique employed favors the isolation of the most common and abundant fungi (often referred as "generalists") (Jeewon and Hyde 2007). The saprophytic condition of the Cerrado's isolates reinforces their ability to use lignocellulose as carbon source, however not all of the isolated fungi were able to secrete significant amounts of cellulase activities for biotechnological use.

The strain L04 identified as *Trichoderma harzianum*, when grown on SCB presented the best ratios of cellulase production among all analyzed strains including *T. reesei* Rut C30. *T. harzianum* are frequently reported as control agent against fungal pathogens (Arantes and Saddler 2010; Banerjee et al. 2010). However, recent studies have also revealed the potential of this fungus for cellulase production and industrial applications (Ahmed et al. 2009; Castro et al. 2010a, 2010b). Likewise, *T. harzianum* has also become a promising system for xylanase production under appropriate conditions (Franco et al. 2004).

Castro et al. (2010b) had previously shown that *T. harzianum* IOC3844 exhibited an expressive production of endoglucanase activity with a fast kinetics with the exponential production phase detected between 31 and 72 hours of fermentation after a short lag phase. In this work, under the same assay conditions, strain L04 showed a shorter lag phase with an earlier exponential production phase starting before 24 hours when grown on SCB. Our results have shown that, unlike the other fungi analyzed in this work, L04 showed a shorter acclimation period when cultivated in a complex substrate such as SCB. This observation is particularly interesting when compared to *T. reesei* Rut C30, a widely used filamentous fungus strain used for the production of cellulolytic enzymes (Peterson and Nevalainen 2012). *T. reesei* Rut C30 shows a longer acclimation to lignocellulosic feedstock and is known to have a better performance for cellulase production when grown on pure cellulosic substrates then on lignocellulosic ones (Juhász et al. 2005), data confirmed by Castro et al. (2010c) where maximum values of volumetric productivity for *T. reesei* Rut C30 were obtained at 333 hours culture.

Strain L04 produced more xylanase, endoglucanase and β-glucosidase activities on different substrates irrespective of the carbon source (Table 2). The best results were obtained when L04 was grown on SCB. Partially delignified cellulignin (PDC) showed to be a poor inducer of L04 cellulases with delayed enzymatic production, contrary to expectations where delignification can improve enzyme production, since the large amount of lignin in SCB could irreversibly adsorb the enzymes produced during fungal cultivation. This would also be expected during hydrolysis where the absence of lignin probably reduces the adsorption of cellulolytic enzymes onto the lignin fraction of biomass (Berlin et al. 2005). This behavior was also reported for a *T. harzianum* strain isolated from the Amazon rainforest grown on pretreated sugarcane bagasse (Delabona et al. 2012).

The electrophoretic profile of L04 using SCB as substrate is similar to the one observed by Silva et al. (2012) using *T. harzianum* strain T4 cultivated in medium containing sugarcane bagasse. These authors showed by different electrophoretic techniques that *T. harzianum* was able to secret active multi enzymatic complexes with cellulolytic and xylanolytic activities which match the high molecular mass signal observed in L04 (Figure 2, lane 5). Both zymograms using CMC and xylan as substrate showed a signal around 20 kDa (Figure 2, lanes 3 and 5). The weaker signal in the xylan zymogram indicates the presence of a ~20 kDa xylanolytic activity which is compatible with other studies of this species that describe xylanase activities around this molecular mass with the lack of cellulase activity (Rezende et al. 2002; Lee et al. 2009; do Vale et al. 2012). The L04 cellulase activity around 20 kDa could correspond to the endoglucanase (EGIII) from *T. harzianum* IOC3844 characterized by Generoso et al. (2012) described with a low molecular mass, lack the cellulose binding domain (CBD) and able to degrade amorphous cellulose such as CMC.

In summary, L04 showed an interesting ability of producing significant yields of cellulase in a short culture time when grown on SCB. Also, it revealed a well-balanced cellulolytic complex, presenting fast kinetics for production of endoglucanases, exoglucanases and β-glucosidases. About 60% glucose yields were obtained from SCB and PDC after 18 hours of hydrolysis. We propose that *T. harzianum* L04 should be considered as a potential candidate for on-site enzyme production using *in natura* sugarcane bagasse as carbon source, in ready supply in a bio-ethanol production plant.

REFERENCES

1. Ahmed S, Bashir A, Saleem H, Saadia M, Jamil A (2009) Production and purification of cellulose-degrading enzymes from a filamentous fungus Trichoderma harzianum. Pak J Bot 41:1411-1419
2. Andreaus J, Filho EXF, Bom EPS (2008) Biotechnology of holocellulose–degrading enzymes. In: Hou CT, Shaw JR (eds) Biocatalysis and Bioenergy, New Jersey, USA: Copyright. pp 197-229
3. Arantes V, Saddler JN (2010) Access to cellulose limits the efficiency of enzymatic hydrolysis: the hole of amorphogenesis. Biotechnol Biofuels 3:4
4. Bailey MJ, Biely P, Poutanen K (1992) Interlaboratory testing of methods for assay of xylanase activity. J Biotechnol 23:257-270
5. Banerjee G, Scott-Craig JS, Walton JD (2010) Improving enzymes for biomass conversion: a basic research perspective. Bioenergy Res 3:82-92
6. Berlin A, Gilkes N, Kilburn D, Bura R, Markov A, Skomarovsky A, Okunev O, Gusakov A, Gregg D, Sinitsyn A, Saddler J (2005) Evaluation of novel fungal cellulase preparations for ability to hydrolyze softwood substrates–evidence for the role of accessory enzymes. Enzym Microb Tech 37:175-184
7. Betancur GJV, Pereira N Jr (2010) Sugarcane bagasse as feedstock for second generation ethanol production. Part I: diluted acid pretreatment optimization. Electron J Biotechnol 13:01-08
8. Bradford MM (1976) A rapid and sensitive method for the quantitation of microgram quantities of protein utilizing the principle of protein-dye binding. Anal Biochem 72:248-254
9. Canilha L, Chandel AK, Milessi TSS, Antunes FAF, Freitas WLC, Felipe MGA, Silva SS (2012) Bioconversion of sugarcane biomass into ethanol: an overview about composition, pretreatment methods, detoxification of hydrolysates, enzymatic saccharification, and ethanol fermentation. J Biomed Biotechnol 2012:989572
10. Castro AM, Carvalho MLA, Leite SGF, Pereira N Jr (2010) Cellulases from Penicillium funiculosum: production, properties and application to cellulose hydrolysis. J Ind Microbiol Biotechnol 37:151-158
11. Castro AM, Ferreira MC, Cruz JC, Pedro KCNR, Carvalho DF, Leite SGF, Pereira N Jr (2010) High-yield endoglucanase production by Trichoderma harzianum IOC-3844 cultivated in pretreated sugarcane mill byproduct. Enzyme Res 2010:854526
12. Castro AM, Pedro K, da Cruz J, Ferreira M, Ferreira MC, Leite SGF, Pereira N Jr (2010) Trichoderma harzianum IOC-4038: a promising strain for the production of a cellulolytic complex with significant beta-glucosidase activity from sugarcane bagasse cellulignin. Appl Biochem Biotechnol 162:2111-2122

13. Delabona PS, Farinas CS, Silva MR, Azzoni SF, Pradella JGC (2012) Use of a new Trichoderma harzianum strain from the Amazon rainforest with preteated sugar cane bagasse for on-site cellulase production. Bioresour Technol 107:517-521

14. do Vale LHF, Gómez-Mendonza DP, Kim MS, Pandey A, Ricart CAO, Filho EXF, Sousa M (2012) Secretome analysis of the fungus Trichoderma harzianum grown on cellulose. Proteomics 12:2716-2728

15. Druzhinina IS, Kopchinskiy AG, Komón M, Bissett J, Szakacs G, Kubicek CP (2005) An oligonucleotide barcode for species identification in Trichoderma and Hypocrea. Fungal Genet Biol 42:813-828

16. Ewing B, Hillier L, Wendl M, Green P (1998) Basecalling of automated sequencer traces using PHRED. I. Accuracy assessment. Genome Res 8:175-185

17. FAOSTAT (2012) Available from: http://faostat.fao.org. Accessed October 5, 2012

18. Franco P, Ferreira H, Ferreira E (2004) Production and characterization of hemicellulase activities from Trichoderma harzianum strain T4. Biotechnol Appl Biochem 40:255-259

19. Generoso WC, Malago-Jr W, Pereira N Jr, Henrique-Silva F (2012) Recombinant expression and characterization of an endoglucanase III (cel12a) from Trichoderma harzianum (Hypocreaceae) in the yeast Pichia pastoris. Genet Mol Res 11:1544-1557

20. Ghose TK (1987) Measurement of cellulase activities. Pure Appl Chem 59:257-268

21. Himmel ME, Ruth MF, Wyman CE (1999) Cellulase for commodity products from cellulosic biomass. Curr Opin Biotechnol 10:358-364

22. Jeewon R, Hyde KD (2007) Detection and diversity of fungi from environmental samples: traditional versus molecular approaches. In: Varma A, Oelmüller R (eds) Soil biology: advanced techniques in soil microbiology, vol. 11, Berlin Heidelberg: Springer-Verlag.

23. Juhász T, Szengyel Z, Réczey K, Siika-Aho M, Viikari L (2005) Characterization of cellulases and hemicellulases produced by Trichoderma reesei on various carbon sources. Process Biochem 40:3519-3525

24. Kumar R, Singh S, Singh OV (2008) Bioconversion of lignocellulosic biomass: biochemical and molecular perspectives. J Ind Microbiol Biotechnol 35:377-391

25. Lee JM, Shin JW, Nam JK, Choi JY, Han IS, Nam SW, Choi YJ, Chung DK (2009) Molecular cloning and expression of the Trichoderma harzianum C4 endo-β-1,4-xylanasa gene in Saccharomyces cerevisiae. J Microbiol Biotechnol 19:823-828

26. Maeda RN, Serpa VI, Rocha VAL, Mesquita RAA, Anna LMMS, Castro AM, Driemeier CE, Pereira N Jr, Polikarpov I (2011) Enzymatic hydrolysis of pretreated sugar cane bagasse using Penicillium funiculosum and Trichoderma harzianum cellulases. Process Biochem 46:1196-1201

27. Mandels M, Weber J (1969) The production of cellulases. Adv Chem Ser 95:391-414

28. Miller GL (1959) Use of dinitrosalicylic acid reagent for determination of reducing sugar. Anal Chem 31:426-428

29. Montenecout BS, Eveleigh DE (1977) Preparation of mutants of Trichoderma reesei with enhanced cellulase production. Appl Environ Microbiol 34:777-782

30. Moreira LRS, Ferreira GV, Santos SST, Ribeiro APS, Siqueira FG, Filho EXF (2012) The hydrolysis of agro-industrial residues by holocellulose-degrading enzymes. Braz J Microbiol 43:498-505

31. Myers N, Mittermeier RA, Mittermeier CG, Fonseca GAB, Kent J (2000) Biodiversity hotspots for conservation priorities. Nature 403:853-858

32. Parachin NS, Siqueira S, de Faria FP, Torres FAG, de Moraes LMP (2009) Xylanases from Cryptococcus flavus isolate I-11: enzymatic profile, isolation and heterologous expression of CfXYN1 in Saccharomyces cerevisiae. J Mol Catal B Enzym 59:52-57

33. Peterson R, Nevalainen H (2012) Trichoderma reesei RUT-C30–thirty years of strain improvement. Microbiology 158:58-68

34. Rezende MI, Barbosa AM, Vasconcelos AFD, Endo AS (2002) Xylanase production by Trichoderma harzianum rifai by solid state fermentation on sugarcane bagasse. Braz J Microbiol 33:67-72

35. Roeder V, Broda P (1987) Rapid preparation of DNA from filamentous fungi. Lett Appl Microbiol 1:17-20

36. Ruegger MJS, Tauk-Tornisielo SM (2004) Atividade da celulase de fungos isolados do solo da Estação Ecológica de Juréia-Itatins, São Paulo, Brasil. Rev Bras Bot 27:205-211

37. Sambrook J, Russel DW (2001) Molecular cloning–a laboratory manual. New York: Cold Spring Harbor Laboratory Press.

38. Silva AJ, Gómez-Mendonza DP, Junquira M, Domoni GB, Ximenes-Filho E, Sousa MV, Ricart CAO (2012) Blue native-PAGE analysis of Trichoderma harzianum secretome reveals cellulases and hemicellulases working as multienzymatic complexes. Proteomics 12:2729-2738

39. Siqueira FG, Siqueira AG, Siqueira EG, Carvalho MA, Peretti BMP, Jaramilho PMD, Teixeira RSS, Dias ES, Félix CR, Filho EXF (2010) Evaluation of holocellulase production by plant-degrading fungi grown on agro-industrial residues. Biodegradation 21:815-824

40. Sun X, Liu Z, Zheng K, Song X, Qu Y (2008) The composition of basal and induced cellulase systems in Penicillium decumbens under induction or repression conditions. Enz Microb Technol 42:560-567

41. Szijártó N, Szengyel Z, Lidén G, Réczey K (2004) Dynamics of cellulase production by glucose grown cultures of Trichoderma reesei Rut-C30 as a response to addition of cellulose. Appl Biochem Biotechnol 113:115-124

42. Tamura K, Dudley J, Nei M, Kumar S (2007) MEGA4: molecular evolutionary genetics analysis (MEGA) software version 4.0. Mol Biol Evol 24:1596-1599

43. Tolan JS (2002) Iogen's process for producing ethanol from cellulosic biomass. Clean Technol Envir 3:339-345

44. White TJ, Bruns T, Lee S, Taylor J (1990) Amplification and direct sequencing of fungal ribosomal RNA genes for phylogenetics. In: Innis MA, Gelfand DH, Sninsky JJ, White TJ (eds) PCR Protocols: a guide to methods and applications, New York, USA: Academic Press. pp 315-322

CONVERSION OF C$_6$ AND C$_5$ SUGARS IN UNDETOXIFIED WET EXPLODED BAGASSE HYDROLYSATES USING *SCHEFFERSOMYCES (PICHIA) STIPITIS* CBS6054

RAJIB BISWAS, HINRICH UELLENDAHL, AND BIRGITTE K. AHRING

9.1 INTRODUCTION

In recent years, ethanol production from renewable sources has received increased attention in a world of dwindling fossil fuels reserves along with the environmental concerns. Commercial production of bio-ethanol is mostly driven by starch- or sucrose-containing feedstocks such as corn, sugarcane, wheat by fermentation with *Saccharomyces cerevisiae* (Wheals et al. 1999). Non-food feedstocks, however, such as lignocellulosic materials including agricultural wastes such as bagasse hold significant potential and have been identified as suitable feedstock sources for ethanol production (Lynd et al. 1991). Lignocellulose based ethanol processes require pretreatment as a first step followed by enzymatic hydrolysis of carbohydrates (Ahring et al. 1996; Margeot et al. 2009). Unlike the hydrolysis of starch- and sugar-based feedstock that results primarily in hexoses, lignocellulose is composed of cellulose and hemicellulose, resulting in both hexoses (C_6) and pentoses (C_5) sugars (Rubin 2008). An efficient pretreatment strategy along with the fermentation of C_6 and C_5 sugars are keys to bring cellulosic ethanol to commercial reality.

Sugarcane bagasse (SCB), the residual plant material of sugarcane, is one of the most abundant lignocellulosic feedstocks suitable for ethanol production (Cardona et al. 2010; Pandey et al. 2000). In addition, its on-site availability at sugarcane-based ethanol process plants is advantageous for large-scale processing. Currently the bagasse generated after sucrose extraction from sugarcane is incinerated to power the plant operation (Shi et al. 2012). SCB is primarily composed of cellulose (40-45%), hemicelluloses (30-35%) and lignin (20-30%) (Cardona et al. 2010). Cellulose is a D-glucose polymer while hemicellulose predominantly consists of D-xylose, a five-carbon sugar (Girio et al. 2010; Jeffries et al. 2007; Skoog and Hahn-Hägerdal 1990). An appropriate pretreatment is essential for efficient enzymatic saccharification (Ahring et al. 1996). Various pretreatment methods have shown the potential to disrupt the cell wall structure of SCB to facilitate the enzymatic hydrolysis of the polysaccharides (Cardona et al. 2010; Martin et al. 2007). Wet explosion is a thermochemical pretreatment method, where biomass is treated at high temperature and pressure. Typically an oxidizing agent such as elemental oxygen or H_2O_2 is added to help disrupt the cell wall structure, and solubilize hemicellulose and lignin. The process is terminated by sudden pressure release to a subsequent flash tank (Ahring and Munck 2006; Rana et al. 2012). In previous studies, the potential of wet explosion pretreatment of bagasse to facilitate saccharification at low enzyme dosage was demonstrated (Biswas et al. unpublished). The oxidative pretreatment strategy was found to improve the cellulose conversion to glucose in the subsequent enzymatic hydrolysis, as well as producing high xylose yields through solubilization of hemicellulose. However, during the processing of hydrolysate for subsequent microbial fermentation, degradation products such as acetate, 5-hydroxymethylfurfural

(HMF), furfural will be formed to various degree known to inhibit the microbial growth and product yields at higher concentration (Bellido et al. 2011; Nigam 2001a; Palmqvist and Hahn-Hägerdal 2000).

The importance of utilizing all hydrolyzed sugar monomers into ethanol for improving process economics is self-evident. *Saccharomyces cerevisiae* is the most commonly used yeast for industrial ethanol fermentation, only capable of glucose fermentation. Some naturally occurring yeast such as *Scheffersomyces stipitis*, *Candida shehatae*, and *Pachysolen tannophilus* are able to ferment both hexoses and pentoses to ethanol. Among the xylose fermenting yeasts, *Scheffersomyces stipitis* seems to be the most promising strain for industrial application due to its high ethanol yield. In addition, this organism is able to ferment most of the sugars glucose, xylose, mannose, galactose and cellobiose (Agbogbo and Coward-Kelly 2008). However, previous studies have shown arabinose is only utilized by *S. stipitis* for cell growth but not for ethanol production (Nigam 2001b). Furthermore, *S. stipitis* also has the natural ability to metabolize some of the sugar degradation compounds present in the hydrolysate after pretreatment (Almeida et al. 2008; Wan et al. 2012). The sensitivity of *Scheffersomyces stipitis* to inhibitors found in lignocellulose hydrolysate has been reported elsewhere (Bellido et al. 2011; Delgenes et al. 1996).

Inhibitory compounds, such as acetic acid, HMF and furfural are produced in different concentrations depending on the pretreatment severity and can inhibit the growth of yeast cell and thus lower the yield and productivity of ethanol fermentation. It was previously reported that prolonged incubation helps to acclimatize *Scheffersomyces stipitis* to these toxic compounds (Delgenes et al. 1996). In the present study, we investigated conversion of both hexose and pentose sugars in the enzymatic hydrolysates of wet exploded sugarcane bagasse without detoxification of the inhibitors to study cell growth and ethanol yields by *S. stipitis* CBS6054. We further compared the cell growth and yields using bagasse xylose hydrolysate containing only xylose with lower concentrations of the inhibitors. The kinetics of cell growth in the hydrolysates compared to synthetic media was also assessed.

9.2 MATERIALS AND METHODS

9.2.1 YEAST STRAIN AND INOCULUM PREPARATION

S. stipitis CBS6054 was obtained from the American Type Culture Collection (ATCC) and was preserved at −80°C in the Bioproducts, Sciences and Engineering Laboratory (BSEL), Washington State University (WSU), USA. The organism was cultivated in a media previously described elsewhere (Agbogbo and Wenger 2006, 2007). A mixture of yeast extract, urea, peptone and xylose (YUPX) in the respective proportions of 1.7, 2.27, 6.65 and 20.0 g/l was filter sterilized (0.22 μm) and used as source of nutrient. 250 ml sterilized Erlenmeyer baffled flasks were used and inoculation was done aseptically. The inoculated medium was incubated in a shake incubator (The Lab Companion IS-971 (R/RF) Floor Model Incubated Shaker, GMI Inc., USA) at 30°C and agitation speed of 140 rpm for 48 h. Microaerobic conditions were maintained by using foam plugs on the Erlenmeyer flasks (Identi-Plugs®, Jaece Industries, Inc., NY). *S. stipitis* cells were harvested towards the end of the exponential growth phase by centrifugation at relative centrifugal force (RCF) 3824 × g for 10 minutes. The harvested cells were washed twice and resuspended in sterilized distilled water in the desired cell concentration and served as inoculum.

9.2.2 WET EXPLOSION PRETREATMENT

Wet explosion pretreatment was performed using the WSU pretreatment pilot plant for disrupting the lignocellulosic matrix and fractioning the lignin and hemicellulosic components as previously described (Rana et al. 2012). Sugarcane bagasse was added to the 10 l pretreatment reactor as wet slurry with 16% dry matter concentration, containing 640 g of oven dried bagasse and 3343 g of tap water. The reactor was hermetically closed, 6 bar of O2 was then purged into the reactor with the headspace of 6 l and the reactor was heated to the desired temperature. Reaction time was 10 minutes at the desired temperature and pressure. Three suitable pretreatment conditions were chosen based on preliminary results on enzymatic hydrolysis of wet exploded bagasse (Table 1). Higher enzyme efficiency and recovery of both glucose and xylose were obtained under condition B followed by condition C, while condition A was found suitable for especially xylose recovery and formation of inhibitors such as weak acid is minimal. Therefore, condition A was chosen for a control condition to obtain hydrolysate contained mostly xylose.

9.2.3 PREPARATION OF HYDROLYSATE FROM WET EXPLODED BAGASSE

9.2.3.1 XYLOSE HYDROLYSATE AFTER SSF

A liquid fraction (AX) containing mostly xylose as fermentable sugar was obtained after simultaneous saccharification and fermentation (SSF) of wet exploded bagasse at condition A (Tables 1 and 2). *Saccharomyces cerevisiae* was used for removing the fermentable glucose for an incubation period of 162 hours. Same enzyme loading of 12.4 mg enzyme protein (EP)/g cellulose at $10.1 \pm 0.1\%$ dry matter was used for the SSF. Since only glucose is utilized by the strain, the remaining liquid fraction after the SSF contained mostly xylose as fermentable sugar. After the fermentation was completed, ethanol produced during SSF was removed by vacuum distillation and the liquid fraction rich in xylose (A_X) was separated for further use.

TABLE 1: Wet explosion pretreatment conditions applied on sugarcane bagasse with a treatment time of 10 minutes.

Pretreatment	Temperature °C	O_2 used (bar)	pH		Dry matter, %	
			Initial	Final	Initial	Final
A	170	6	5.85	3.12	16.0	15.5
B	185	6	5.85	3.05	16.0	16.2
C	200	6	5.85	2.93	16.0	14.0

TABLE 2: Composition (g/l) of the substrates used for fermentation by *Scheffersomyces (Pichia) stipitis.*

Substrate	Initial sugar concentration		Initial inhibitor concentration		
	Glucose	Xylose	Acetic acid	HMF	Furfural
A_X^a	0.0	14.7± 0.0	1.0±0.0	1.2± 0.0	0.4±0.0
B_{GX}^b	17.3± 0.5	9.6± 0.2	3.2±0.1	0.4± 0.0	0.5±0.0
C_{GX}^c	42.8± 0.8	6.3± 0.0	6.9±0.1	1.2± 0.0	0.8±0.0
S_{GX}^d	6.1±0.0	15.2± 0.0	0.0	0.0	0.0
S_G^d	27.2± 0.5	0.0	0.0	0.0	0.0
S_X^d	0.0	25.6± 0.4	0.0	0.0	0.0

[a] hydrolysate after pretreatment at condition A and the simultaneous saccharification and fermentation (SSF) using Saccharomyces cerevisiae.
[b] hydrolysate after pretreatment at condition B and enzymatic hydrolysis.
[c] hydrolysate after pretreatment at condition C and enzymatic hydrolysis.
[d] respective commercial sugar (granular powder) was used as control substrate, Fisher Chemical, USA.

9.2.3.2 HYDROLYSATE WITH MIXED SUGARS AFTER ENZYMATIC HYDROLYSIS

After pretreatments under condition B and C (Table 1), enzymatic hydrolysis was carried out on the whole wet exploded material (slurry) without any solid–liquid separation. For saccharification, a mixture of the two commercial enzymes Cellic® CTec2 and Cellic® HTec2 (Novozymes, USA) were used in a ratio of 85:15 (%, v/v), respectively, with the enzyme loading of 12.4 enzyme protein (EP)/g cellulose at 10.1±0.1% dry matter. The enzyme protein (EP) content of Cellic® CTec2 and Cellic® HTec2 determined prior to enzymatic hydrolysis were 279±8 and 251±12 mg EP/ml, respectively. Enzymatic hydrolysate BGX and CGX were obtained from enzymatic hydrolysis of the pretreated samples under condition B and C, respectively (Table 2). Hydrolysates were always filter sterilized (0.2 μm, Millipore, USA) prior to inoculation.

9.2.3.2.1 SHAKE FLASK FERMENTATION

Shake flask fermentation was conducted in duplicates with the hydroly-sates (Table 2) under same conditions as previously described. Filter ster-ilized synthetic medium S_{GX}, S_G and S_X were prepared using commercial sugar(s) (Fisher Chemical, USA) contained glucose + xylose, glucose and xylose, respectively, with the concentration as depicted in Table 2. Erlen-meyer baffled flasks were used with a volume of 50 ml. Adjustment of pH to 6.0 ± 0.5 was performed for hydrolysates with 1 M NaOH whenever this was needed to ensure a pH of at least 6.0. Each flask contained 30 ml of hydrolysate or sugars solution (glucose and/or xylose in DI water), 1 ml of nutrient solution and 1 ml of inoculum (initial cell concentration 1 g/l). Nutrient solution was prepared by dissolving 4.25 g of yeast extract, 5.68 g of urea and 16.40 g of peptone in 23.68 ml of water to reach a volume of 50 ml. All fermentation flasks were supplemented with sufficient carbon sources (i.e., hydrolysate or commercial sugar) and nutrients to produce equivalent amount of cell mass and to exhibit similar growth rates under the favorable conditions ensured. The flasks were incubated for 106 hours except for hydrolysate CGX which was incubated for 174 hours. 2 ml of sample was withdrawn after 0, 6, 12, 24, 36, 48, 58, 82, 106 and 174 hours (in case of hydrolysate CGX) for analysis of sugar and inhibitor concen-trations, cell concentration and pH.

9.2.3.2.2 ANALYTICAL METHODS

Cell concentrations were determined by optical density (OD) measure-ment of the cells using spectrophotometer (Jenway 6405 UV/Visible, NJ, USA) system at 600 nm (1 OD = 0.17 g/l of dry cells). Glucose, xy-lose, arabinose, acetic acid, ethanol, HMF and furfural were quantified by HPLC on an Aminex HPX-87H column (Bio-Rad, Hercules, USA) at 60°C with 4 mM H_2SO_4 as an eluent with a flow rate of 0.6 ml/min. HPLC was equipped with refractive index and UV visible detector. All samples were filtered through a 0.45 μm PTFE membrane (Acrodisc® Syringe Fil-ters, 13 mm, Pall® Life Sciences, USA) prior to HPLC analysis. The pH was monitored using InLab® Micro combination pH electrode (precision ± 0.001 pH).

9.3 RESULTS

9.3.1 EFFECTS OF INHIBITORS ON SUGAR UTILIZATION AND ETHANOL YIELDS

The main parameters measured for the fermentation by *Scheffersomyces stipitis* CBS6054 on the different hydrolysates and control media are displayed in Table 3. Sugar utilization, ethanol yield, inhibitor concentration, pH and growth kinetics of hydrolysates B_{GX} and C_{GX} are presented in Figures 1 and 2, respectively.

The sugar utilization ratio and ethanol yield ($Y_{p/s}$) ranged from 88–100% and 0.33–0.41±0.02 g/g, respectively, in all the hydrolysates and controls tested. The ethanol yields ($Y_{p/s}$) of hydrolysate A_X, B_{GX} and C_{GX} were 0.41±0.02, 0.39±0.02 and 0.38±0.02 g/g, respectively. Ethanol yields were higher when using hydrolysates after pretreatment than control substrates, i.e., commercial sugars (Table 3).

The fermentation of xylose alone, after pretreatment at condition A (170°C, 6 bar O_2) and SSF, took 58 h to convert 100% sugar (Table 3), which is longer than that of mixed sugars in the hydrolysate B_{GX} after pretreatment condition B (185°C, 6 bar O_2), which took 36 h (Figure 1A). Both glucose and xylose were converted for the hydrolysates B_{GX} and C_{GX} obtained after the pretreatment and enzymatic hydrolysis of SCB, containing inhibitors in comparatively higher concentrations among others. Fermentation of enzymatic hydrolysate C_{GX} after pretreatment at condition C (200°C, 6 bar O_2) resulted in a prolonged fermentation time of 82 h with initial lag phase of 12 h (Figure 2A). The delay in sugar conversion is likely due to the presence of inhibitors such as acetate, HMF and furfural at the concentrations of 6.9±0.1, 1.2 and 0.8 g/l, respectively. Nonetheless, ethanol concentration was found to be 18.7±1.1 g/l after 82 h of incubation. While *S. stipitis* adapted to the inhibitors, the fermentation was completed with an ethanol yield of

0.38 ± 0.02 g/g at 82 h. Although the utilization of sugars was limited to 88% within this time, sugar conversion was found to be 95% after 174 h of fermentation.

Taking into consideration that no detoxification was performed except the adjustment of pH with NaOH to 6.0 ± 0.5, it was found that the fermentation was only inhibited in bagasse hydrolysate C_{GX} after pretreatment at condition C (200°C, 6 bar O_2). Acetic acid was converted in all fermentation experiments especially with hydrolysate B_{GX} and C_{GX} resulting an increase in pH (Agbogbo and Wenger 2007). After 82 h of fermentation, 100% acetic acid was metabolized in hydrolysate B_{GX} (Figure 1B). Hence, for the hydrolysate C_{GX}, it took 174 h to bring the acetic acid concentration to 1.3 g/l from 6.9 ± 0.1 g/l (Figure 2B). Moreover, both HMF and furfural were utilized by *S. stipitis* CBS6054 within the first 12 hours of fermentation for hydrolysate B_{GX} and C_{GX}.

9.3.2 EFFECTS OF INHIBITORS ON CELL GROWTH

When comparing the growth kinetics of *Scheffersomyces stipitis* CBS6054 in Figures 1C and 2C, the initial cell concentration of 1 g/l increased for all hydrolysates and grew to various final cell concentrations on the different hydrolysate medium. The highest amount of cell mass (g/l) produced in mixed sugars hydrolysate B_{GX} after 106 h of incubation was 4.02 ± 0.02, while 3.34 ± 0.02 and 3.52 ± 0.09 in hydrolysate A_X and hydrolysate C_{GX}, respectively (Table 3).

Cell mass production was higher in all hydrolysates than found in synthetic medium (S_{GX}, S_G and S_X). Exponential growth was observed for hydrolysate AX and BGX (Figure 1C) during the initial 48 h without any noticeable lag phase. Cell mass in hydrolysate A_X and B_{GX} after 48 h were measured to 2.81 and 3.52 g/l, respectively. On the other hand, no cell growth was observed in hydrolysate C_{GX} within the first 12 h (Figure 2C).

The highest cell growth rate of 0.079 g/l/h was found in hydrolysate B_{GX} followed by 0.064 g/l/h in synthetic media S_G (Table 3). Acetic acid concentrations in the hydrolysates A_X, B_{GX} and C_{GX} were 1.0 ± 0.0, 3.2 ± 0.1 and 6.9 ± 0.1 g/l, respectively (Table 2).

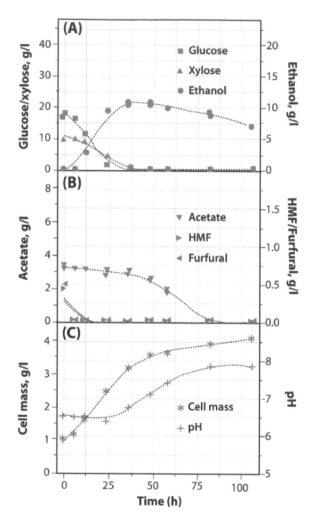

FIGURE 1: Fermentation results for hydrolysate B_{GX} obtained after enzymatic hydrolysis of wet exploded bagasse under condition at 185°C with 6 bar O_2; (A) sugar conversion and ethanol production; (B) conversion of inhibitors; and (C) cell growth and pH in batch fermentation with *Scheffersomyces (Pichia) stipitis* CBS6054.

9.4 DISCUSSION

To realize the industrial ethanol production from hydrolysis of pretreated lignocellulose, it is essential to obtain strains capable of converting all

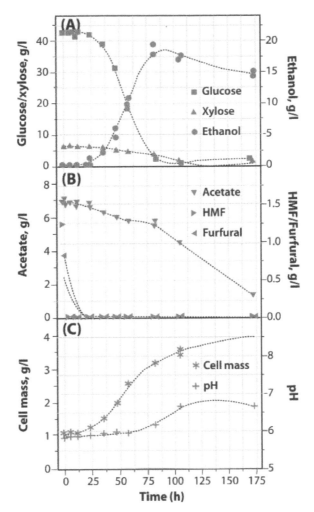

FIGURE 2: Fermentation results for hydrolysate C_{GX} obtained after enzymatic hydrolysis of wet exploded bagasse under condition at 200°C with 6 bar O_2; (A) Sugar conversion and ethanol production; (B) conversion of inhibitors; and (C) cell growth and pH in batch fermentation with *Scheffersomyces (Pichia) stipitis* CBS6054.

the major sugars as well as being able to cope with the inhibitors present as sugar degradation product in the hydrolysate. Our present work demonstrates that the native strain *Scheffersomyces (Pichia) stipitis* CBS6054

TABLE 3: Summaries of fermentation results at highest ethanol concentration time points using *Scheffersomyces (Pichia) stipitis.*

Substrate	Fermentation time, (f)h	Ethanol, g/l	Sugar utilized, %	$Y_{p/s}$*, g/g	Cell growth rate, g/l/h	Cell mass, g/l at (f)h
A_X	58	6.1±0.0	100±0.0	0.41± 0.02	0.043	2.81±0.02
B_{GX}	36	10.4±0.2	100±0.0	0.39± 0.02	0.079	3.31±0.00
C_{GX}	82	18.7±1.1	88±0.0	0.38± 0.02	0.049	3.16±0.00
S_{GX}	76	8.2±0.0	100±0.0	0.39± 0.00	0.045	2.48±0.00
S_G	36	10.1±0.1	99±0.8	0.37± 0.00	0.064	2.33±0.05
S_X	82	8.5±0.2	100±0.0	0.33± 0.01	0.051	2.81±0.03

* Yp/s = ethanol yield coefficient, was calculated as the grams of ethanol produced per grams of sugar converted.

is suitable for ethanol fermentation of both glucose and xylose present in hydrolysates of wet exploded bagasse without the need for detoxification, achieving substantial ethanol yields. The ethanol yield from xylose in the hydrolysate after pretreatment at 170°C with 6 bar O_2 and SSF was 0.41±0.02 g/g while a yield of 0.39±0.02 g/g was achieved for the fermentation of glucose and xylose in the hydrolysate after pretreatment at 185°C with 6 bar O_2 and enzymatic hydrolysis of wet exploded bagasse. The yields are in agreement with the results found in corn stover hemicellulose hydrolysate with similar inhibitor concentrations using *Scheffersomyces (Pichia) stipitis* CBS6054 (Agbogbo and Wenger 2007). Our results are comparable to those observed with adapted *S. stipitis* strains (Nigam 2001a,b). The utilization of glucose was more rapid than for xylose in the different hydrolysates. This similar observation in assimilation of sugars has been reported elsewhere (Agbogbo and Wenger 2007; Bellido et al. 2011; Nigam 2001a). In the presence of both glucose and xylose (B_{GX}, C_{GX}), conversion of glucose started prior to xylose conversion. In mixed substrate fermentation, significant xylose utilization is initiated by *Scheffersomyces (Pichia) stipites* once glucose concentration in the medium is below 20 g/l (Agbogbo et al. 2006).

Conversion of glucose and xylose was not completely inhibited for the hydrolysates B$_{GX}$ and C$_{GX}$, in the presence of known inhibitors such as acetate, HMF and furfural. Our study shows that the favorable growth condition for cell mass production is likely due to the mixed sugars, where glucose is converted more readily than xylose. Our results compare favorably with previous reports on fermentation of sugarcane bagasse hydrolysate (Rudolf et al. 2008). In contrast, Bellido et al. (2011) found that xylose was not utilized in 168 h of fermentation experiments using *Scheffersomyces (Pichia) stipitis* DSM3651 on filtered hydrolysate of steam exploded wheat straw using the whole slurry with acetate, HMF and furfural concentrations at 1.52, 0.05 and 0.14 g/l, respectively. Acetic acid is released from the esterified form of arabinoxylans during the processing of lignocellulose hydrolysate. The cleavage of the acetyl group occurs when lignocellulose undergoes high temperature, oxidation treatment and even in enzymatic hydrolysis process we further see a liberation of acetic acid. Previous studies showed the yeast cell growth is inhibited at an acetic acid concentration of about 2–5 g/l (Bellido et al. 2011; Nigam 2001a). Acetic acid is a weak acid having high pKa value of 4.75 (25°C) at zero ionic strength. pKa value refers to the pH value at which buffering capacity of the acid is highest and the concentration of dissociated and undissociated form of the acid are equal (Palmqvist and Hahn-Hägerdal 2000). The risk of inhibition due to liposoluble diffusion of undissociated weak acid across the plasma membrane can be reduced by increasing the pH (Palmqvist and Hahn-Hägerdal 2000). Therefore, favorable pH for the fermentation of the hydrolysates containing acetic acid will be between 5.5 and 6.5. Our study suggests that acetic acid can be utilized by *S. stipitis* as a substrate at a lower concentration that may not be inhibitory for cell growth at starting pH between 6.0 and 6.5. A similar observation of acetic acid conversion by *Scheffersomyces (Pichia) stipitis* was also reported (Agbogbo and Wenger 2007) during fermentation of corn stover hydrolysate. The product formed from acetic acid metabolism by *S. stipitis* CBS6054 is unknown. HMF and furfural are produced during the processing of hydrolysate, by degradation of hexose and pentose sugars, respectively. Apparently, the tested concentration levels of HMF and furfural were not affecting the fermentation and growth of *S. stipitis* CBS6054. Yeasts including *S. stipitis* can metabolize furfural to furfuryl alcohol and the enzyme NADH- dependent yeast alcohol dehydrogenase (ADH) is responsible for the reduction (Huang et al. 2009). In the present investigation, HMF and furfural were completely metabolized by the strain before significant utilization of sugars started. This was also previously reported by others (Almeida et al. 2008; Wan et al. 2012) and indicates that S. stipitis CBS6054 is readily capable of converting HMF and furfural in the tested lignocellulose hydrolysate from sugarcane bagasse. Cell growth was highest (0.079 g/l/h) in hydrolysate containing mixed sugars and inhibitors such as acetate, HMF and furfural at concentrations of 3.2 ± 0.1, 0.4 and 0.5 g/l, respectively, indicating that the processing of bagasse hydrolysate under this condition will not inhibit the growth of S. stipitis.

A lag phase of 12 hours is observed in the fermentation of C_{GX} hydrolysate. This lag phase is possibly due to a higher concentration of inhibitor in hydrolysate C_{GX} such as acetate (6.9 ± 0.1 g/l), HMF (1.2 g/l) and furfural (0.8 g/l). Similar observation was also reported by others (Agbogbo and Wenger 2007; Sreenath and Jeffries 2000). Although S. stipitis exhibited prolonged fermentation time for the hydrolysate processed at 200°C with 6 bar O_2 containing the inhibitors at higher concentration, ethanol concentration up to 18.7 ± 1.1 g/l was obtained with an ethanol yield of 0.38 ± 0.02 g/g after 82 h. However, after adaptation to the hydrolysate C_{GX} within 12 h, exponential growth was observed. The performance was significantly improved shortly after 12 h of incubation. This lag phase can be overcome in a continuous process using initial high cell density and also by recycling the cells adapted to the inhibitors (Bellido et al. 2011).

REFERENCES

1. Agbogbo F, Coward-Kelly G: Cellulosic ethanol production using the naturally occurring xylose-fermenting yeast, Pichia stipitis . Biotechnol Lett 2008, 30:1515–1524.
2. Agbogbo F, Wenger K: Effect of pretreatment chemicals on xylose fermentation by Pichia stipitis . Biotechnol Lett 2006, 28:2065–2069.
3. Agbogbo F, Wenger K: Production of ethanol from corn stover hemicellulose hydrolyzate using Pichia stipitis . J Ind Microbiol Biot 2007, 34:723–727.
4. Agbogbo FK, Coward-Kelly G, Torry-Smith M, Wenger KS: Fermentation of glucose/xylose mixtures using Pichia stipitis . Process Biochem 2006, 41:2333–2336.

5. Ahring B, Munck J: Method for treating biomass and organic waste with the purpose of generating desired biologically based products. 2006. Patent. WO 2006/032282 A1

6. Ahring B, Jensen K, Nielsen P, Bjerre A, Schmidt A: Pretreatment of wheat straw and conversion of xylose and xylan to ethanol by thermophilic anaerobic bacteria. Bioresource Technol 1996, 58:107–113.

7. Almeida J, Modig T, Röder A, Lidén G, Gorwa-Grauslund M: Pichia stipitis xylose reductase helps detoxifying lignocellulosic hydrolysate by reducing 5-hydroxymethyl-furfural (HMF). Biotechnol Biofuels 2008, 1:12.

8. Bellido C, Bolado S, Coca M, Lucas S, González-Benito G, García-Cubero MT: Effect of inhibitors formed during wheat straw pretreatment on ethanol fermentation by Pichia stipitis . Bioresource Technol 2011, 102:10,868–10,874.

9. Cardona C, Quintero J, Paz I: Production of bioethanol from sugarcane bagasse: status and perspectives. Bioresource Technol 2010, 101:4754–4766.

10. Delgenes J, Moletta R, Navarro J: Effects of lignocellulose degradation products on ethanol fermentations of glucose and xylose by Saccharomyces cerevisiae , Zymomonas mobilis , Pichia stipitis , and Candida shehatae . Enzyme Microb Tech 1996, 19:220–225.

11. Girio F, Fonseca C, Carvalheiro F, Duarte L, Marques S, Bogel-Lukasik R: Hemicelluloses for fuel ethanol: a review. Bioresource Technol 2010, 101:4775–4800.

12. Huang C, Lin T, Guo G, Hwang W: Enhanced ethanol production by fermentation of rice straw hydrolysate without detoxification using a newly adapted strain of pichia stipitis . Bioresource Technol 2009, 100:3914–3920.

13. Jeffries T, Grigoriev I, Grimwood J, Laplaza J, Aerts A, Salamov A, Schmutz J, Lindquist E, Dehal P, Shapiro H, Jin Y, Passoth V, Richardson P: Genome sequence of the lignocellulose-bioconverting and xylose-fermenting yeast Pichia stipitis . Nat Biotechnol 2007, 25:319–326.

14. Lynd L, Cushman J, Nichols R, Wyman C: Fuel ethanol from cellulosic biomass. Science 1991, 251:1318–1323.

15. Margeot A, Hahn-Hagerdal B, Edlund M, Slade R, Monot F: New improvements for lignocellulosic ethanol. Curr Opin Biotech 2009, 20:372–380.

16. Martin C, Klinke H, Thomsen A: Wet oxidation as a pretreatment method for enhancing the enzymatic convertibility of sugarcane bagasse. Enzyme Microb Tech 2007, 40:426–432.

17. Nigam J: Ethanol production from hardwood spent sulfite liquor using an adapted strain of Pichia stipitis . J Ind Microbiol Biot 2001, 26:145–150.

18. Nigam J: Ethanol production from wheat straw hemicellulose hydrolysate by Pichia stipitis . J Biotechnol 2001, 87:17–27.

19. Palmqvist E, Hahn-Hägerdal B: Fermentation of lignocellulosic hydrolysates. II: inhibitors and mechanisms of inhibition. Bioresource Technol 2000, 74:25–33.

20. Pandey A, Soccol C, Nigam P, Soccol V: Biotechnological potential of agro-industrial residues. I: sugarcane bagasse. Bioresource Technol 2000, 74:69–80.

21. Rana D, Rana V, Ahring BK: Producing high sugar concentrations from loblolly pine using wet explosion pretreatment. Bioresource Technol 2012, 121:61–67.

22. Rubin E: Genomics of cellulosic biofuels. Nature 2008, 454:841–845.

23. Rudolf A, Baudel H, Zacchi G, Hahn-Hägerdal B, Lidén G: Simultaneous saccharification and fermentation of steam-pretreated bagasse using Saccharomyces cerevisiae TMB3400 and Pichia stipitis CBS6054. Biotechnol Bioeng 2008, 99:783–790.
24. Shi J, Yang Q, Lin L, Zhuang J, Pang C, Xie T, Liu Y: The structural changes of the bagasse hemicelluloses during the cooking process involving active oxygen and solid alkali. Carbohyd Res 2012, 359:65–69.
25. Skoog K, Hahn-Hägerdal B: Effect of oxygenation on xylose fermentation by Pichia stipitis . Appl Environ Microb 1990, 56:3389–3394.
26. Sreenath H, Jeffries T: Production of ethanol from wood hydrolyzate by yeasts. Bioresource Technol 2000, 72:253–260.
27. Wan P, Zhai D, Wang Z, Yang X, Tian S: Ethanol Production from Nondetoxified Dilute-Acid Lignocellulosic Hydrolysate by Cocultures of Saccharomyces cerevisiae Y5 and Pichia stipitis CBS6054. Biotechnol Res Int 2012, 1–6.
28. Wheals A, Basso L, Alves D, Amorim H: Fuel ethanol after 25 years. Trends Biotechnol 1999, 17:482–487.

PART III

ECONOMIC AND ENVIRONMENTAL FACTORS

CHAPTER 10

BIOELECTRICITY VERSUS BIOETHANOL FROM SUGARCANE BAGASSE: IS IT WORTH BEING FLEXIBLE?

FELIPE F. FURLAN, RENATO TONON FILHO, FABIO H. P. B. PINTO, CALIANE B. B. COSTA, ANTONIO J. G. CRUZ, RAQUEL L. C. GIORDANO AND ROBERTO C. GIORDANO

10.1 INTRODUCTION

It is already a consensus that a shift of the global energy matrix towards renewable sources is mandatory. Yet, the role that each specific alternative will play, say, at the year 2050, will be defined along the road, depending on technological developments, political options by stakeholders, economical and social demands. Anyway, in this scenario ethanol will certainly be an important biofuel.

Sugarcane is known to be the most efficient crop for 1G ethanol production, with an energy balance of 9.3 produced/consumed tonne of oil equivalent (toe) [1]. During the 1970's the Brazilian government initiated the National Ethanol Program (PROALCOOL, in Portuguese, [2]) to decrease national dependence on oil. Since then, the use of 1G ethanol as a vehicle fuel has been consolidated, and presently 86% of the cars sold in this country are flex-fuel, running with any mixture of ethanol and gasoline [3]. In modern facilities, ethanol production is a highly integrated process, with sugarcane bagasse burnt in boilers to supply the industrial plant energy demands, further exporting the surplus of electric energy to the grid.

One of the alternatives for the industrial production of 2G ethanol is the biochemical route, i.e., acid or enzymatic hydrolysis of the biomass followed by fermentation of the resulting sugars. Logistics and transportation of the lignocellulosic raw material may be a bottleneck for 2G ethanol [4]. From this point of view, sugarcane bagasse has an important advantage, since it has already been collected and processed for the extraction of the juice, being immediately available at the plant site. Moreover, sugarcane trash (mostly the leaves) can be transported with the stalk, after small adaptations of the mechanical harvesting – although part of this biomass must be left for covering the fields [5]. Since the process must be energetically self-sufficient, the addition of sugarcane trash as boiler fuel can increase the amount of bagasse available for hydrolysis, therefore enhancing ethanol yields.

Industrial production of 2G ethanol is still not consolidated in large scale. Therefore, an economic analysis is important to indicate if it is the most interesting alternative, specially when compared to selling electric energy (bioelectricity). Nevertheless, this answer is not unique, given the volatility of relative prices: biomass electricity prices in public auctions in Brazil ranged from 85.35 USD/MWh to 53.02 USD/MWh in the last two auctions (Aug/2010 and Aug/2011) [6], a -37.9% variation. The spot market (that buys surplus power, beyond the amount contracted during the auctions) presented an even higher range of prices, between 3.26 USD/MWh and 341.13 USD/MWh over the last nine years (Jan/2003 - Dez/2012) [7]. Ethanol prices are equally volatile, changing from 0.258 USD/L to 0.818 USD/L (for the hydrated fuel) over the same period of nine years [8]. All prices above, used in this study, were calculated in Brazilian reais, brought to December/2012 value (to take into account the inflation in the period), and converted to US dollars using the exchange rate value of 2.077 BRL/USD (dez/2012).

Sugarcane biorefineries have been intensively studied by the recent literature. Seabra et al. (2011) [9] presented economic and environmental analyses of a sugarcane biorefinery. The authors concluded that, although electric energy presented a better economic feasibility, its environmental impact was greater than the one for second generation ethanol. An economic analysis comparing 2G ethanol production with electric energy was also done by Dias et al. (2011) [10]. The authors concluded that, although for the present technology electric energy is a better option, 2G ethanol can compete with it if sugarcane trash is used, provided that new technologies could increase yields. On the other hand, Macrelli et al. (2012) [11] presented results for several sugarcane biorefineries configurations and concluded that 2G ethanol from sugarcane is already competitive with 1G starch-based ethanol in Europe. The advantages of the integrated production of 1G and 2G ethanol production based on sugarcane was highlighted by Dias et al. (2012) [12]. The integrated biorefinery presented higher ethanol production rates, and better economic and environmental performance, when compared to a stand-alone 2G ethanol-from-sugarcane bagasse plant.

The 2G biorefinery could be flexible, just as the industry that employs current 1G technology already is, shifting between sugar and ethanol production. This new flexible biorefinery might be able to choose between electric energy and 2G ethanol production. The present study focuses on assessing the economic feasibility of a flexible biorefinery, for an autonomous distillery (not considering the manufacture of sugar), and comparing it to the dedicated 1G + electric energy and 1G + 2G ethanol biorefineries. The process chosen as case study uses enzymatic hydrolysis of the biomass (sugarcane bagasse) and ethanolic fermentation of hexoses and pentoses. Specifically, this study presents a computational applicative that may be a useful tool for the process scheduling of future cane-based biorefineries but, beyond that scope, that may also support decision-making concerning national energy policies. Such computationally robust tool was developed within an equation-oriented simulator (EMSO) [13] and is based on phenomenological modelling, at least for the most important unit operations and reactors that are present in the process.

10.2 RESULTS AND DISCUSSION

10.2.1 PROCESS SIMULATION

Two boundary process configurations were considered and compared: an industry producing 1G ethanol and burning all sugarcane bagasse and 50% of the trash produced in the field for power generation in a Rankine cicle (BioEE) and another one using all bagasse surplus (the biorefinery must be energetically self-sufficient) for 2G ethanol production, integrated to the 1G facility (BioEth). The most important information for BioEE and BioEth streams (as enumerated in Figure 1 and Figure 2) is presented in Table 1 and Table 2. 74% of the bagasse could be diverted to 2G ethanol in BioEth. This accounted for an increase in ethanol production of 25.8% when compared to BioEE. At this condition, a specific production yield for 2G ethanol of 120.7 L/tonne of bagasse was obtained, which is a conservative estimate based on current yields (158 L/tonne of lignocellulosic material [12]). The increase in steam consumption for 2G ethanol was entirely fulfilled by the burning of lignin and of non-hydrolyzed cellulose. Since 65% of the cellulose was hydrolyzed, 35% of the material was still able to be separated and used as fuel. Table 3 shows ethanol production (total and specific) for both cases.

As shown in Table 4, the steam demand increased by 56.3% from BioEE to BioEth. This represented a steam consumption of 8.8 kg of steam/L of 2G ethanol, compared to 4.0 kg of steam/L of 1G ethanol. This higher consumption was mainly caused by the low concentration of both glucose in the hydrolyzed liquor (9 wt% compared to 17.7 wt% for sugarcane juice) and of ethanol in the C5 wine (21.3 g/L compared to 78.8 g/L for C6 wine). Nevertheless, the higher energy demand of 2G ethanol was diluted by the 1G's, and the overall specific steam consumption was 5.0 kg of steam/L of ethanol (1G + 2G).

Table 5 presents the energy demand by plant sector. Since not all electric power produced was consumed in BioEth, it delivered electric energy to the grid, too. It is clear that the impact of the production of 2G ethanol on the overall energy demand was low. The major impact, in fact, was on the condensing turbine, since all bagasse surplus was diverted to 2G ethanol production. Therefore, this turbine was absent in the BioEth plant.

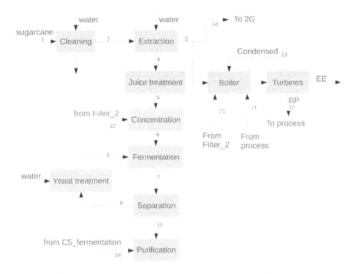

FIGURE 1: Process diagram for first generation ethanol production.

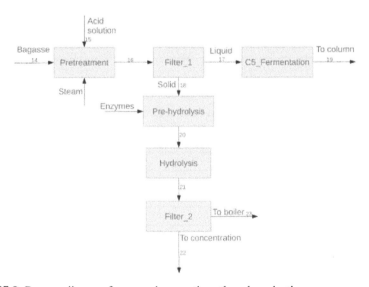

FIGURE 2: Process diagram for second generation ethanol production.

TABLE 1: Summaries of fermentation results at highest ethanol concentration time points using *Scheffersomyces (Pichia) stipitis*.

Stream n°	Mass flow (kg/h)	Temperature (°C)	Pressure (bar)	Fraction of sugars
1	500000	30	1	0.145
2	495675	30	1	0.144
3	131791	30	1	0.02
4	512584	30	1	0.134
5	513565	90	1	0.133
6	389697	111.7	1	0.177
7	578699	31	1	0
8	90016	31	1	0
9	189003	30	1	0
10	488683	31	1	0
11	299182	110.5	2.5	0
12	293196	189.3	2.5	0
13	193026	55.7	0.1	0

TABLE 2: Summaries of fermentation results at highest ethanol concentration time points using *Scheffersomyces (Pichia) stipitis*.

Stream n°	Mass flow (kg/h)	Temperature (°C)	Pressure (bar)	Fraction of sugars
1	500000	30	1	0.145
2	495675	30	1	0.144
3	131791	30	1	0.02
4	512584	30	1	0.134
5	513565	90	1	0.133
6	332396	111.7	1	0.206
7	493607	31	1	0
8	76920.5	31	1	0
9	161212	30	1	0
10	416687	31	1	0
11	196577	110.5	2.5	0
12	188785	189.3	2.5	0
13	0	55.7	0.1	0
14	97613.3	30	1	0.02
15	174018	30	1	0

TABLE 2: CONTINUED.

16	271631	120	2	0.02
17	201087	100	1	0.02
18	70544	100	1	0.01
19	201045	27	1	0.02
20	153036	50	1	0.02
21	153036	50	1	0.08
22	104925	50	1	0.09
23	48111	50	1	0.05

TABLE 3: First and second generation ethanol production rates.

	BioEE	BioEth
Ethanol production (L/h)	45796.6	57580.8
Specific ethanol production (1G + 2G)	91.6	115.2
(L/tonne of sugarcane (TC))		
2G ethanol production (L/h)	0	11784.2
Specific 2G ethanol production	0	120.7
(L/tonne of bagasse (TB))		

Plant capacity: 500 tonnes of sugarcane per hour.

TABLE: Steam consumption (total and specific).

Sector	Steam consumption (total (kg/h) / specific (kg/TC))	
	BioEE	BioEth
Juice treatment	51971/103.9 [a]	51971/103.9 [a]
Concentration	185240/370.5	241210/482.6
Distillation	121775/243.5 [a]	188455/376.9 [b]
Pretreatment*	0	38493/394.3
Total	185240/370.5	289550/579.1

*Steam specific consumption in kg/TB.
[a] Uses steam from concentration step.
[b] Uses both steam from concentration step and from the back pressure turbine (5.2% of the latter).
Plant capacity: 500 tonnes of sugarcane per hour.

TABLE 5: Power consumption divided by sector (positive values for produced energy and negative for consumed).

Sector	Power demand/production (kW)	
	BioEE	BioEth
Mills	-6368.7	-6368.7
Pretreatment	0	-62.0
Hydrolysis	0	-246.4
Centrifuges	-380.9	-599.6
Pumps	-1389.8	-1055.9
Back pressure turbine	+32106.7	+49863.9
Condensing turbine	+48051.6	0
Total	+72019	+41531.3

10.2.2 ECONOMIC ANALYSIS

An economic analysis was performed for both process configurations described above (BioEE and BioEth). Besides, a flexible biorefinery (Flex), which can switch between cogeneration and 2G ethanol, was also considered. This option might enable a better exploitation of the seasonality of both ethanol and electric energy prices (Table 6).

Since the condensing turbine is not necessary for BioEth and, moreover, steam production in this case was lower than in BioEE, the investment in the combined heat and power plant decreased from BioEE to BioEth. On the other hand, an increase in costs for fermentation, distillation and tankage was necessary in BioEth to account for the higher ethanol production. For the flexible biorefinery, it must be suitable for both maximum ethanol and maximum electric energy production, which makes its investment costs the highest. Table 7 presents the investment costs for the cases considered.

TABLE 6: Ethanol and electric energy average seasonality over the period 2003-2012.

Month	Ethanol (%)	Electric energy (spot market) (%)
January	11.19%	22.97%
February	7.64%	-24.71%
March	9.29%	-27.39%
April	5.76%	-31.52%
May	-8.90%	-26.25%
June	-12.61%	-16.08%
July	-8.79%	-5.77%
August	-6.53%	-11.77%
September	-5.37%	20.39%
October	-1.11%	40.86%
November	2.81%	41.92%
December	6.60%	17.36%

Percentage variation from the average mean.

TABLE 7: Investment costs by sector of the biorefinery, internal rate of return and net present value.

Sector	Cost (10^6 USD)		
	BioEE	BioEth	Flex
Sugarcane reception, preparation and milling	38.5	38.5	38.5
Combined heat and power plant	50.2	42.9	50.2
Fermentation, distillation and tankage	30.8	35.4	35.4
Sugarcane juice treatment	23.1	23.1	23.1
Piping, general tankage and valves	15.4	15.4	15.4
Licenses, project and ground leveling	7.7	7.7	7.7
2G (pre-treatment, hydrolysis and C5 fermentation)	0	9.6	9.6
Total	165.8	172.7	180.0
IRR*	7.6%	8.3%	8.0%
NPV (10^6 USD)*	-34.5	-30.0	-41.8

*for an ethanol price of 513.7 USD/m^3.

The flexible biorefinery (Flex) allows the decision (considered here to be in a monthly basis) to operate between the two boundary cases represented by BioEE and BioEth. Table 8 shows the chosen option (between electric energy or 2G ethanol) over the whole period considered. It is worth mentioning that both earnings and costs were equally distributed through the year for the flexible biorefinery. Therefore, ethanol and/or bagasse must be stocked to assure the selling during the off-season period. As reported by Agblevor et al. [14], bagasse composition is not greatly affected by storage period, even when it is exposed to weather conditions. The authors verified that only the upper third and the outer region of the dry interior were attacked by micro-organisms and had their composition changed in a period of 26 weeks.

The results of the economic analysis are presented in Table 7. Both IRR and Net Present Value (NPV) showed similar results, with BioEth being the best option, followed by the flexible biorefinery. As it can be seen, none of the options presented a positive NPV or, equivalently, an IRR higher than the Minimum Acceptable Rate of Return (MARR), assumed to be 11%/yr, when the present market prices for ethanol and electric energy in Brazil are considered. Nevertheless, as it will be shown later in the sensitivity analysis, an increase in ethanol price of 21.1% can turn all options feasible. Besides, specifically for the BioEth biorefinery, an increase of only 11.5% in this price would assure feasibility.

Unfortunately, a direct comparison of the obtained result with the ones from literature is not a straightforward task. There is a large variability in the economic premises and in the technical solutions for the biorefineries that can be considered in a techno-economic study. For example, Seabra and Macedo (2011) [9] considered a 2G biorefinery adjacent to the 1G industrial plant (i.e., not sharing utilities and equipment, just importing bagasse), while Dias et al. (2011) [10] and Macrelli et al. (2012) [11] considered that 1G and 2G production plants were integrated (in different degrees). Even when similar processes are considered, the results can be quite different: while the autonomous 1G industrial plant in Dias et al. (2011) [10] obtained an IRR of 15.9 %, in Macrelli et al. (2012) [11] a similar plant obtained an IRR of 32.1 %.

10.2.3 SENSITIVITY ANALYSIS

Since ethanol prices presented an approximately normal distribution, its standard deviation was calculated and a variation equivalent to one standard deviation (20.5%) was used for the sensitivity analysis. On the other hand, electric energy price (spot market) did not present a normal distribution and its influence was better seen when the double of its current price was considered. Therefore, for this case a variation of 40% was considered and, for the other ones that had only punctual data available, a variation of 20% was chosen, thus assuming a percent variation similar to the one in ethanol prices. For Flex sensitivity analysis, the variations described were applied before the simulation of the seasonality effect on the prices.

Figure 3 presents the sensitivity analysis of electric energy selling prices in public auctions on the IRR. As expected, the influence was higher for BioEE, since it has the higher amount of electric energy being sold in

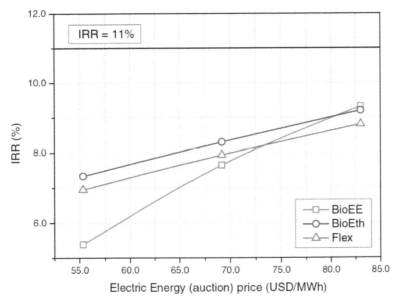

FIGURE 3: Impact of electric energy selling prices (annual auctions) on the internal rate of return. All other prices kept unchanged.

public auctions. BioEth and Flex produced the same amount of electric energy to be sold in the auction market. Therefore, both were equally influenced by the variation in auction prices. As it can be seen, the effect of the electric energy price on the biorefineries economic performance was not strong enough to make them feasible. The cash flow provided by the selling of electric energy was one order of magnitude smaller than the one for ethanol.

Electric energy prices in the spot market only influenced Flex's IRR (Figure 4). The sensitivity to this parameter was quite small, though. As seen in Table 8, only a few months were dedicated to electric energy production, even when the spot market price was assumed equal to twice its current value. The income from electric energy sold in the spot market was small, compared to the one coming from public auctions. It should be mentioned that an increase in both auction and spot prices can be expected in the near future in Brazil. Thermoelectric plants using natural gas currently complement the production of hydroelectric energy during the dry period, and this is hardly sustainable, both in economic and environmental perspectives. Therefore, an increase in these prices, associated with improvements in the distribution grid, could stimulate investments in cogeneration.

It is worth noticing that while Flex could reproduce almost perfectly the behaviour of BioEth (except for the higher investment cost), the same was not true for BioEE. This is due to the fact that the latter sold all its available electric energy in the auction market, which pays higher prices (in general), while the flexible biorefinery would not do this.

The sensitivity with respect to enzyme prices was not as significant as initially expected (Figure 5). Although enzyme costs played a major role in composing 2G ethanol prices (75% of the total cost), this influence was diluted by the 1G ethanol costs. Therefore, the overall ethanol costs (1G + 2G) changed from 416.3 USD/m³ (for an enzyme price of 2.02 USD/kg) to 392.6 USD/m³ (for an enzyme price of 1.35 USD/kg), a variation of 5.9%, for a 40% reduction in enzyme price.

Figure 6 shows the impact of ethanol selling prices on the IRR for the current price for electric energy in spot market (a) and for a price equal to twice its current value (b). It is clear that ethanol price presented the higher influence for all biorefineries. In fact, it was the only factor strong enough

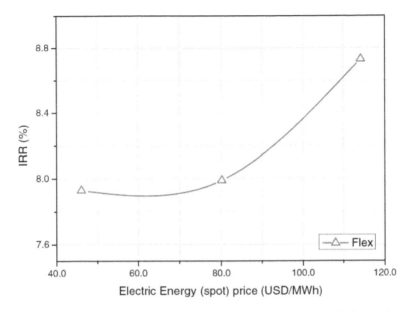

FIGURE 4: Impact of electric energy selling prices (spot market) on the internal rate of return. The central point corresponds to a price (80.2 USD/MWh) equal to twice the current value. All other prices kept unchanged.

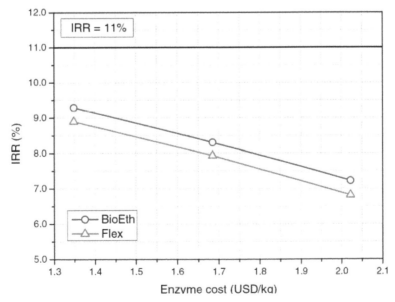

FIGURE 5: Impact of enzyme costs on the internal rate of return. All other prices kept unchanged.

TABLE 8: Chosen option between electric energy surplus (EE) and 2G ethanol production (2G) for the flexible biorefinery.

Month	1st to 8th year	9th to 13th year	14th to 25th year
January	2G	2G	2G
February	2G	2G	2G
March	2G	2G	2G
April	2G	2G	2G
May	2G	2G	2G
June	2G	2G	2G
July	2G	2G	2G
August	2G	2G	2G
September	EE	2G	2G
October	EE	EE	2G
November	EE	2G	2G
December	2G	2G	2G

The numbers in the first row are the interval of years for which the behavior of the flexible biorefinery remained constant. Spot energy price assumed to be equal to twice its current value (80.2 USD/MWh).

to make the biorefineries economically feasible, within the range spanned in this study. Accordingly, BioEE, BioEth and Flex became feasible for an ethanol price of 622.1 USD/m³, 572.8 USD/m³ and 583.1 USD/m³, respectively. This means an increase in ethanol price of 21.1%, 11.5% and 13.5%, respectively. While in Figure 6 (a) it is clear that BioEth was always superior to the flexible biorefinery, when a higher price in the spot market was considered (Figure 6 (b)) the flexible biorefinery became less influenced by negative variations of the ethanol price. The switch towards electric energy production for the flexible biorefinery is clear in this case. If the current ethanol price were considered, no extra electric energy was produced in any month for the whole period considered (25 years). On the other hand, when ethanol price is on its lower value, the flexible bio-refinery switched to electric energy production during 30% of the months, becoming superior to BioEth.

The non-linear behaviour of the system becomes evident in Figure 6. This is mostly caused by the fact that both tax rates and dividends only apply when there is profit. Therefore, the positive influence of an increase in ethanol prices on the IRR is attenuated by both of them and the second derivative of these curves is negative.

Since there are many uncertainties regarding the investment costs on the second generation ethanol production process, a sensitivity analysis was also performed for this parameter. Maximum and minimum values of twice and half the base investment cost for the 2G ethanol process were considered. As seen in Figure 7, the impact of the investment cost on the IRR is small. The decrease in IRR between the maximum and minimum investment costs was only of -0.72% and -0.68% for BioEth and Flex, respectively. Therefore, it is expected that the uncertainty of this value will not invalidate the analyses performed.

It is worth highlighting that the 2G process is coupled to the 1G process, which is responsible for the low investment cost of the former, when compared to literature [15], because costs of fermentation, distillation and combined heat and power stages are included in our 1G cost, which is scaled proportionally to the combined process flows.

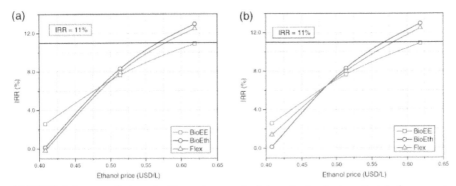

FIGURE 6: Impact of ethanol selling prices on the internal rate of return. In (a), the current spot energy price (40.1 USD/MWh) is considered, while in (b) a spot energy price of twice this value (80.2 USD/MWh) is used. All other prices kept unchanged.

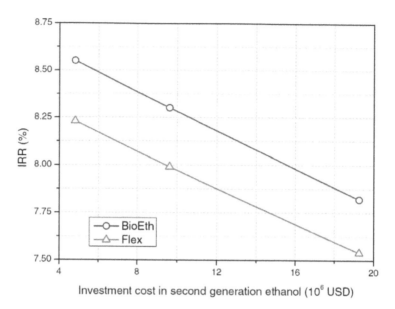

FIGURE 7: Impact of the investment cost on second generation ethanol on the internal rate of return. All other prices kept unchanged.

10.3 CONCLUSIONS

A flexible sugarcane biorefinery (Flex) was simulated and compared to dedicated first generation + cogeneration (BioEE) and first + second generation ethanol (BioEth) biorefineries. The flexible one presented an inferior economic performance in all cases for 2012 market prices. Nevertheless, if an increase in electric energy prices in the spot market were considered, the flexible biorefinery could be the best option.

In general, all biorefineries were not economically feasible for 2012 selling prices and costs. This conclusion was actually validated in practice by recent governmental actions (April/2013) which aimed to improve the competitiveness of the ethanol industry. Additionally, of all parameters considered in the sensitivity analysis, ethanol prices were the only ones that could make the biorefineries economically viable, within the studied range of values. In fact, an increase in ethanol price of 21.1% would be sufficient to make feasible all biorefineries. Particularly for BioEth, a 11.5% increase in ethanol prices would be enough for viability.

Enzyme prices, on the other hand, were less significant than it could be expected. This is due to the fact that 2G ethanol costs were diluted by 1G ethanol's, produced in higher volumetric rates. Therefore, the overall ethanol production cost in the integrated plant was not greatly influenced by enzyme prices.

Finally, it is obvious that the quantitative results presented here are dependent on the economic scenario proposed and on the assumed process yields and energy demands, but the presented methodology is general.

10.4 METHODS

10.4.1 PROCESS IMPLEMENTED

The 1G ethanol production process was simulated based on a typical industrial plant. Figure 1 shows the main required operations. First, sugarcane is cleaned with water to remove dirt carried during harvesting. Next, the sugars are extracted by mechanical pressure. The solution containing the extracted sucrose (juice) follows a series of steps in order to remove impurities which could decrease fermentation yields. The solution is concentrated and fermented by *Saccharomyces cerevisiae*, producing an alcoholic solution which is purified in distillation columns, producing hydrous ethanol. Table 9 shows the main parameters used in the simulations. A rigorous description of the main models used in this study can be found in Furlan et al.[16].

The cogeneration system was also considered in the simulations. It uses sugarcane bagasse, sugarcane trash, and alternatively, non-hydrolyzed cellulose and lignin, as fuel and produces steam and electric energy to supply process demands using a Rankine cicle. It was considered that 50% of sugarcane trash is brought from the field to be burnt in the boiler. Since around 140 kg of trash (dry basis) is produced per tonne of sugarcane [9,17], it was considered that a flow of 35 tonnes of sugarcane trash/h was fed to the boiler. If there is a surplus of electric energy, it can be sold to the grid. The cogeneration system includes a boiler, a back-pressure turbine and a condensing one. Table 10 presents the main data used in the simulation of the cogeneration system.

TABLE 9: Main data for first generation ethanol production.

Input	Value	Unit
Sugarcane flow	500	tonne/h
Sugarcane TRS (Total Reducing Sugars)	15.86	% w/w
Cleaning section		
Sugar losses	1.5	%
Cleaning efficiency	70	%
Water flow	1	kg/kg of sugarcane
Sugar extraction section		
Sugarcane bagasse humidity	50	% w/w
Sugar recovery (first mill)	70	%
Sugar recovery (total)	96	%
Duty	16	kWh/tonne of fiber
Water flow	30	% w/w
Sugarcane juice treatment section		
CaO flow	2	kg/tonne of juice
CaO concentration	10	% w/w
Heating final temperature	105	°C
Steam used	53698	kg/h
Water losses in flash	6495.5	kg/h
Polymer	3534	kg/h
Polymer concentration	0.05	% w/w
Sugar losses (decanter)	6.8	%
Sludge humidity	50	% w/w
Clarifier temperature (after decanter)	92	°C
Sugar losses (filter)	5.6	%
Filter cake humidity	70	% w/w
Water flow (filter)	116	%
Sugarcane juice concentration section		
Evaporators area	8000	m²
Outlet sugar concentration	21.4	% w/w
Steam consumption	190584	kg/h
Steam produced	181657	kg/h
Pressure of steam produced	2.5	bar
Fermentation section		
Fermentation yield	89	%
Yeast concentration (wine)	14	% w/w
Wine ethanol concentration	9	°GL
Yeast concentration (after separation)	70	% w/w
Ration of yeast rich stream / sugar solution	33	% w/w
Ethanol purification section		
Specific steam consumption (1G)	2.7	kg/L of ethanol
Specific ethanol production (1G)	91.6	L/tonne of sugarcane
Specific vinasse + phlegm production (1G)	10.1	kg/L of ethanol

TABLE 10: Main data for the cogeneration system.

Parameter	Value	Unit
Cellulose LHV[a]	15997.1	kJ/kg
Hemicellulose LHV[a]	16443.3	kJ/kg
Lignin LHV[a]	24170	kJ/kg
Boiler outlet vapor pressure	65.7	bar
Boiler outlet vapor temperature	520	°C
Boiler efficiency	92	%
Back-pressure turbine outlet pressure	2.5	bar
Back-pressure turbine efficiency	68	%
Condensing turbine efficiency	70	%

[a]Lower Heating Value (LHV) calculated using data from Wooley and Putsche [18].

2G ethanol was produced via the biochemical route, using weak acid pretreatment and enzymatic hydrolysis. The main steps can be seen in Figure 2. First, bagasse is pretreated with a solution of H_2SO_4 (3 wt% at 120°C and 2 bar of pressure). At this point, most hemicellulose is hydrolyzed, increasing cellulose accessibility. A filter (Filter_1) is used to separate the solid fraction from the liquid. The solid fraction is pre-hydrolyzed in a horizontal reactor, in order to decrease mixing power demands and water usage. The second hydrolysis is carried out in a stirred reactor without any further addition of water or enzymes. The solid fraction (non-hydrolyzed cellulose + lignin) is separated from the glucose solution in a filter and sent to the boiler to increase steam production. On the other hand, the liquid fraction is directed to the concentration step, being mixed to the 1G juice. The liquid fraction from Filter_1 is sent to a (SIF) reactor, where the xylose in the solution is transformed to xylulose and fermented by *Saccharomices cerevisiae* [19]. The resulting alcoholic solution is sent to the distillation columns with the wine from hexose fermentation. The parameters used for this section are shown in Table 11.

TABLE 11: Main data for second generation ethanol production.

Main data used in the simulation	Value	Unit
Pretreatment		
Pressure	2	Bar
Temperature	121	°C
Cellulose to glucose conversion	8.0	%
Hemicellulose to xylose conversion	74.0	%
Solid/liquid ratio	0.2	
Acid solution concentration	3	wt%
Volumetric power (mixing)[a]	342	W/m³
Space-time	40	min
Reactor volume	182	m³
Pre-hydrolysis		
Cellulose to glucose yield	20	%
Solid/liquid ratio	0.2	
Enzyme/Cellulose ratio	67.34 (20)	g/kg (FPU/g)
Space-time	18	h
Temperature	50	°C
Hydrolysis		
Cellulose to glucose yield	65	%, w/w
Solid/liquid ratio	0.178	
Volumetric power (mixing) [a]	302.5	W/m³
Enzyme/Cellulose ratio	67.34 (20)	g/kg (FPU/g)
Space-time	54	h
Temperature	50	°C
C5 Fermentation		
Xylose to ethanol yield	70	%, w/w
Temperature	30	°C
Space-time	9	h

[a]Calculated using data from Pereira et al. [20].

10.4.2 EMSO SOFTWARE

EMSO [13] was the software chosen as the platform for the simulations in this study. It is an equation-oriented, general purpose process simulator with its own modelling language [21]. Besides the several models for the main process pieces of equipments, the software also allows the user to implement his/hers own models. The software has several numeric solvers for solution of algebraic and differential-algebraic systems, and users can plug in their own numerical routines (in C/C++ or FORTRAN). Physical and thermodynamic properties can be added to the package database by the user whenever needed.

10.4.3 ECONOMIC ANALYSIS

Table 12 shows the main economic premisses. Except for 2G ethanol costs, all process costs were obtained from industry (at Dez/2012). 2G ethanol cost is composed by enzyme prices plus all other costs, which are assumed equivalent to 1G's, due to the lack of industrial information on this topic. Ethanol and electric energy (spot market) selling prices were considered as the mean value over the period between Jan/2003 and Dez/2012. All

TABLE 12: Economic data, base case used as reference.

Process and economic data	Value
Time usage	80%
Days of operation	210 days/year
Ethanol direct/indirect costs (1G)	94.75 USD/m³
Sugarcane costs (1G)	314.78 USD/m³
Ethanol production cost(2G, extra cost)	290.1 USD/m³
Electric energy production cost	38.9 USD/MWh
Ethanol transportation cost	28.9 USD/L
Administrative and general costs	1.1 USD/TC
Ethanol selling price	513.7 USD/m³
Electric energy selling price (public auction)	69.2 USD/MWh
Electric energy selling price (spot market)	40.1 USD/MWh
Enzymes	1.68 USD/kg
Depreciation	10%(p.y.)
Minimum acceptable rate of return	11%(p.y.)
Decrease in production cost due to learning curve*	0.3(1)%(p.y.)
Tax rate (income and social contributions)	34%

*for 1G(2G).

values were adjusted for inflation in Brazil in the period and converted to US dollars using the exchange rate of 2.077 BRL/USD (dez/2012). For the flexible biorefinery, the economic analysis was made in a monthly basis and, for this case, both ethanol and electric energy price variations due to seasonality were considered (Table 6).

10.5 ABBREVIATIONS

1G: First generation; 2G: Second generation; BioEE: First generation biorefinery using all sugarcane bagasse (and trash) to produce electricity in a Rankine cicle; BioEth: First generation biorefinery using sugarcane bagasse surplus to produce second generation ethanol; EE: Electric energy; EMSO: Environment for Modeling Simulation and Optimization; Flex: Flexible biorefinery capable of operating as both BioEE and BioEth; IRR: Internal Rate of Return; MARR: Minimum Acceptable Rate of Return; NPV: Net Present Value; SIF: Simultaneous Isomerization and Fermentation; TB: Tonne of bagasse; TC: Tonne of sugarcane.

REFERENCES

1. Macedo IC, Seabra JEA, Silva JEAR: Green house gases emissions in the production and use of ethanol from sugarcane in Brazil: The 2005/2006 averages and a prediction for 2020. Biomass Bioenergy 2008, 32:582-595.
2. Zanin GM, Santana CC, Bon EPS, Giordano RLC, Moraes FF, Andrietta SR, Neto CCC, Macedo IC, Fo DL, Ramos LP, Fontana J: Brazilian bioethanol program.
3. Appl Biochem Biotechnol 2000, 84:1147-1163.
4. National Association of Motor Vehicles (ANFAVEA): Brazilian automotive industry yearbook. Tech. rep., São Paulo, 2012
5. Gnansounou E: Production and use of lignocellulosic bioethanol in Europe: Current situation and perspectives.
6. Bioresour Technol 2010, 101:4842-4850.
7. Hassuani SJ, Leal MRLV, Macedo IC: Biomass power generation: Sugarcane bagasse and trash. Piracicaba: United Nations Development Programme and Sugarcane Technology Centre; 2005.
8. Electric Energy National Agency – ANEEL (in portuguese) http://www.aneel.gov.br
9. Electric Energy Commercialization Chamber – CCEE (acronym in portuguese) http://www.ccee.org.br/

10. Center of Advanced Studies in Applied Economy – CEPEA/ESALQ/USP (in portuguese) http://www.cepea.esalq.usp.br

11. Seabra JE, Macedo IC: Comparative analysis for power generation and ethanol production from sugarcane residual biomass in Brazil. Energy Policy 2011, 39:421-428. http://www.sciencedirect.com/science/article/pii/S0301421510007706

12. Dias MO, Cunha MP, Jesus CD, Rocha GJ, Pradella JGC, Rossell CE, Maciel Filho R, Bonomi A: Second generation ethanol in Brazil: Can it compete with electricity production? Bioresour Technol 2011, 102(19):8964-8971.

13. Macrelli S, Mogensen J, Zacchi G, et al.: Techno-economic evaluation of 2 nd generation bioethanol production from sugar cane bagasse and leaves integrated with the sugar-based ethanol process. Biotechnol Biofuels 2012, 5:22.

14. Dias M, Junqueira T, Cavalett O, Cunha MP, Jesus C, Rossell C, Filho R, Bonomi A: Integrated versus stand-alone second generation ethanol production from sugarcane bagasse and trash. Bioresour Technol 2012, 103:152-161.

15. Soares RP, Secchi AR: EMSO: A new environment for modelling, simulation and optimisation. Comput Aided Chem Eng 2003, 14:947-952.

16. Agblevor F, Rejai B, Wang D, Wiselogel A, Chum H: blueInfluence of storage conditions on the production of hydrocarbons from herbaceous biomass. Biomass Bioenergy 1994, 7:213-222.

17. Eggeman T, Elander RT: Process and economic analysis of pretreatment technologies. Bioresource Technol 2005, 96(18):2019-2025.

18. Furlan FF, Costa CBB, Fonseca GC, Soares RP, Secchi AR, Cruz AJG, Giordano RC: Assessing the production of first and second generation bioethanol from sugarcane through the integration of global optimization and process detailed modeling. Comput Chem Eng 2012, 43:1-9.

19. Canilha L, Chandel AK, Suzane dos Santos Milessi T, Antunes FAF, Luiz da Costa Freitas W, das Graças Almeida Felipe M, da Silva SS: Bioconversion of sugarcane biomass into ethanol: An overview about composition, pretreatment methods, detoxification of hydrolysates, enzymatic saccharification, and ethanol fermentation. J Biomed Biotechnol 2012, 2012:1-15.

20. Wooley R, Putsche V: Development of an ASPEN PLUS physical property database for biofuels components. NREL, Tech. rep., Report MP-425-20685 1996, 38

21. Silva C, Zangirolami T, Rodrigues J, Matugi K, Giordano R, Giordano R: An innovative biocatalyst for production of ethanol from xylose in a continuous bioreactor. Enzyme Microb Technol 2012, 50:35-42.

22. Pereira LTC, Pereira LTC, Teixeira RSS, Bon EPS, Freitas SP: Sugarcane bagasse enzymatic hydrolysis: rheological data as criteria for impeller selection. J Ind Microbiol Biotechnol 2011, 38(8):901-907.

23. Rodrigues R, Soares RP, Secchi AR: Teaching chemical reaction engineering using EMSO simulator. Comput Appl Eng Educ 2010, 18(4):607-618.

CHAPTER 11

ENVIRONMENTAL ASSESSMENT OF RESIDUES GENERATED AFTER CONSECUTIVE ACID-BASE PRETREATMENT OF SUGARCANE BAGASSE BY ADVANCED OXIDATIVE PROCESS

IVY DOS SANTOS OLIVEIRA, ANUJ K. CHANDEL, MESSIAS BORGES SILVA, AND SILVIO SILVÉRIO DA SILVA

11.1 BACKGROUND

In the last few decades, studies on viable process for second generation ethanol production from lignocellulosic biomass have gained significant momentum worldwide. Deployment of cellulosic ethanol as an alternative of gasoline may provide unique environmental, economic and strategic benefits over to fossil fuels [1,2]. However, concerns like environmental pollution and cost economics of ethanol production technologies are required in-depth analysis for the establishment of biorefineries [3].

Sugarcane bagasse (SB) is the preferred choice of raw material for ethanol production in countries like Brazil, India and China where it is

Environmental Assessment of Residues Generated After Consecutive Acid-Base Pretreatment of Sugarcane Bagasse by Advanced Oxidative Process. © 2013 Oliveira et al. Sustainable Chemical Processes *2013, 1:20 doi:10.1186/2043-7129-1-20; licensee Chemistry Central Ltd. Creative Commons Attribution License (http://creativecommons.org/licenses/by/2.0).*

generated in plentiful amount every year [4,5]. SB like any other ligno-cellulosic material is a complex polymer which is consisted of three major constituents such as cellulose, hemicellulose and lignin. In order to utilize carbohydrate fraction of SB for ethanol production via microbial fermentation, it is necessary to use appropriate pretreatment method for lignin removal. Recently, consecutive acid-base pretreatment process has been found successful for the efficient removal of hemicellulose and lignin leaving cellulose for the cellulolytic enzymes action for its conversion into glucose [6,7].

During the dilute acid pretreatment of SB, hemicellulose is converted into various sugars primarily xylose and some other compounds such as furfurals, phenolics, acids and metals. This acid pretreated SB so called cellulignin when further exposed to dilute sodium hydroxide based pre-treatment, considerable fraction of lignin is removed [8]. Alkali-mediated pretreatment methods degrade the lignin and release phenolic compounds, aromatic alcohols and aldehydes which strongly inhibit the microbial metabolism [9-11]. Compounds derived from hemicellulose and lignin during pretreatment, if discarded in open environment, heavily pollutes the land and water. Furfurals, acids, and phenolic compounds (p-hydroxyben-zoic acid, m-hydroxybenzoic acid, vanillic acid, syringic acid, p-hydroxy-benzaldehyde, vanillin, cinnamic acid, syringaldehyde and others) [12,13] contribute a huge environmental damage due to high toxic content, bioac-cumulation in different food chains even at low concentrations [14].

An efficient treatment method is required to overcome the contamination of phenolic compounds, furans and weak acids. In this line, several methods have been investigated such as adsorption on activated carbon, photo-catalysis using TiO_2, activated carbon post-treatment, anaerobic treatment; autohydrolysis and organosolv process and several types of advanced oxidation processes (Fenton, electro-Fenton, sono-electro-Fenton and photo-electro-Fenton) [15-20]. Amongst them, advanced oxidative process (AOP) has shown promising results for the reduction of total phenolics and organic matter from various waste liquors. AOP are defined as potential processes that are capable of producing hydroxyl radicals ($\bullet OH$), highly oxidative species, in high amounts for mineralizing organic materials to carbon dioxide, water and inorganic ions. Majority of AOP processes are performed at lower temperatures which use energy to produce highly reactive intermediaries with high oxidation or reduction potential. The hydroxyl radicals may be obtained from powerful oxidants, such as H_2O_2 and O_3, combined with irradiation. These processes have shown a great advantage to degrade the pollutants [21-25].

Advancing the AOP process, Fenton reagent has been shown to be very efficient for oxidation of organic compounds which are toxic and non-biodegradable [26]. The Fenton reaction is defined as a catalytic generation of hydroxyl radicals from a chain reaction between ferrous ion (Fe2+) and hydrogen peroxide (H_2O_2) in an acid medium. H_2O_2 is a powerful oxidative agent, when it is catalyzed by ferrous sulfate, it produces the free radical •OH (hydroxyl) which has 60% higher oxidation power than the peroxide. In addition, this radical has the ability of degrading the organic material of effluents in a more efficient way [27]. With a high oxidative potential, these hydroxyl radicals are the responsible for oxidation of organic compounds present in waste waters. The radical Fe2+ is the reaction catalyst. Nowadays, this process is used for treating a wide variety of toxic organic compounds that do not respond to biological treatments, or even in rehabilitation of contaminated land [28,29]. This study aims to evaluate the environmental impact of residues generated during the consecutive acid-base pretreatment of sugarcane bagasse. Advanced oxidative process (AOP) was used based on photo-Fenton reaction mechanism (Fenton Reagent/UV) for the elimination of total phenolics in lignin solution in conjunction with reduction in BOD/COD ratio.

11.2 RESULTS AND DISCUSSION

11.2.1 CHARACTERISTICS OF LIGNIN RESIDUE

The lignin residue was characterized based on the physico-chemical parameters, such as pH, color, TOC, COD, BOD, turbidity, COD/BOD ratio and TP amount. Table 1 shows the physical and chemical characteristics of alkaline hydrolysates (native lignin solution) [30]. The COD/BOD ratio in native lignin solution was 0.03 lower than the standard value (0.2) revealing that the substances present in lignin solution residues are resistant to biological oxidation [31]. The color and turbidity analysis demonstrate

TABLE 1: Physical and chemical characteristics of residual lignin.

Parameters	Characteristics	Standards*
True color (PtCo)	3621.33	Absent
Turbidity (NTU)	50.8	Absent
pH	7 – 8	5 – 9
COD (mg/L O_2)	5870.25	8
BOD (mg/L O_2)	169	60
COD/BOD	0.03	-
TOC (mg/L)	2053.0	-
Total Phenol (mg/L)	10.64	0.5

*Effluents standards in hybrid bodies - Article 18 [31] and [32] respectively. (-) Unspecified.

that the effluent has intense color and turbidity. The presence of phenolic compounds in the alkaline hydrolysates shows the important characteristic of lignin solutions. During the photo-Fenton reaction of lignin solution, extensive amount of foam was generated. Excessive foam causes interference during the photo-Fenton reaction. Therefore, it is necessary to strictly control the foam generation. During the photo-Fenton reaction, all amount of hydrogen peroxide was consumed.

11.2.2 CHANGES IN TOTAL OXYGEN CONCENTRATION (TOC)

Table 2 demonstrates the percentage change in the TOC of lignin solution after photo-Fenton reactions in each experiment carried out as per the L9 orthogonal design of experiments. As can be observed from Table 2, experiment 2 and 9 only have the higher values of standard variation (Si2) showing 3.62 and 3.43 respectively. Different modifications in AOP have shown the satisfactory results for the treatment of the lignin solution. For example, Ninomiya et al. [33] recently studied the sonocatalytic-Fenton reaction for the degradation of lignin. Sonocatalytic-Fenton reaction showed synergistically enhanced •OH radical generation. The •OH radical generation was applied to lignin degradation and biomass pretreatment.

TABLE 2: Profile of percentage reduction of TOC values in all 9 experiments (photo-Fenton reaction) performed as per L_9 Taguchi matrix.

Exp.	TOC range (%)		Average (%)	Si^{2*} (%)
	1° Data set	2° Data set		
1	91.92	93.68	92.80	1.55
2	90.21	92.90	91.56	3.62
3	91.96	93.49	92.73	1.17
4	90.49	90.40	90.45	0.00
5	93.97	94.60	94.28	0.20
6	92.30	92.92	92.61	0.19
7	95.18	96.28	95.73	0.60
8	91.82	93.58	92.70	1.55
9	92.08	94.70	93.39	3.43

$*Si^2$ = variance..

Lignin degradation ratio by sono-catalytic-Fenton reaction was 60.0% at 180 min. Ma et al. [32] studied photo-catalytic degradation of lignin with the use of catalysts $TiO2$ and $Pt/TiO2$. The results showed that application of UV irradiation alone has almost no effect on the reduction of dissolved organic carbon (DOC). However, the addition of $TiO2$ and $Pt/TiO2$ reduced the original DOC (251 mg/L) by more than 40% within 30 minutes of treatment. Makhotkina et al. [34] also used Fenton and H_2O_2 photo-assisted reactions for and observed 85% lignin oxidation, at 0.1 M concentration under UV-radiation at pH 8.3.

Table 3 presents the average percentage change in TOC in all nine experiments carried out according to Taguchi L9 orthogonal design. Experimental run 7 (temperature 35°C; pH 2.5; Fenton concentration 144 mL H_2O_2 + 153 mL Fe^{2+}; UV range 16W) showed the 95.73% TOC reduction. On the other hand, experimental run 4 (temperature 30°C; pH 2.5; Fenton concentration 120 mL H_2O_2 + 120 mL Fe^{2+}; UV range 28W) showed the minimum percentage change in TOC value (90.45%). This positive effect of UV radiation on degradation reaction was attributed to the reduction of Fe^{3+} to Fe^{2+} which in turn reacts with H_2O_2 allowing for the continued Fenton reaction. The absorbance of ferric ions may extend into the visible region, depending on the pH because the pH influences the

TABLE 3: Average value of TOC percentage reduction in lignin solution using photo-Fenton after in all 9 experiments designed as per L_9 Taguchi matrix

Exp.	Temperature (°C)	pH	Fenton (mg L⁻¹)	UV (W)	TOC average (%)
1	1	1	1	1	92.80
2	1	2	2	2	91.56
3	1	3	3	3	92.73
4	2	1	2	3	90.45
5	2	2	3	1	94.28
6	2	3	1	2	92.61
7	3	1	3	2	95.73
8	3	2	1	3	92.70
9	3	3	2	1	93.39

formation of hydroxylated species, which have higher absorption in the visible [35]. It is interesting to note that pH around 3.0 proved to be more effective for the reduction of organic matter present in the residue using photo-Fenton process. Several studies have confirmed this behavior [35-38]. Higher concentration of the Fenton's reagent was more effective for

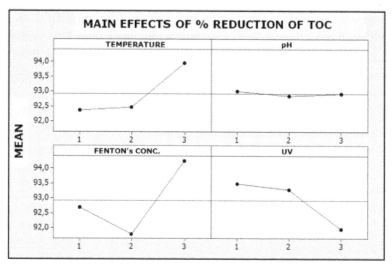

FIGURE 1: Main effects of different process variables (temperature, pH, H_2O_2 concentration and UV radiation) on TOC percentage reduction in lignin solution treated by photo-Fenton process.

TABLE 4: Average value of TOC percentage reduction in lignin solution using photo-Fenton after in all 9 experiments designed as per L_9 Taguchi matrix

Factors	Sum of square	Degree of freedom	Average sum of square	F	p-value
Temperature	9.50	2	4.70	3.50	0.0763
pH	0.10	2	0.03	0.02	0.9766
Fenton Concentration	18.60	2	9.30	6.80	0.0156
Residual	8.40	2	4.20	3.10	0.0962
UV	12.30	9	1.40	-	-

the reduction of organic matter due to the generation of hydroxyl radicals in greater amount. The extensive degradation of lignin was possibly due to the oxidative properties of the Fenton's reagent [39].

The effect of individual parameters on TOC reduction during photo-Fenton reaction has been shown in Figure 1. It is clear from the Figure 1 that pH had no significant effect on the TOC percentage reduction. The Fenton reagent (144 mL of H_2O_2 + 153 mL of Fe^{2+}) and temperature showed high impact in the process. It can be observed that the UV irradiation presence (16 W) or absence has a better contribution in the percentage TOC reduction. One of the possible explanations for this fact is the intense color of the residue, which interferences UV irradiation process. It can also be observed that the higher temperature (35°C) has a great significance (Table 3).

High Fenton reagent level majorly influenced the percentage TOC reduction. Table 4 presents the ANOVA of involved factors in both lignin residue treatment and photo-Fenton process, according to L9 Taguchi experimental matrix. ANOVA presented in Table 4 clearly demonstrates that the Fenton reagent concentration was the most significant effect factor in the TOC percentage reduction, with F equals 6.8 and p-value equals 0.0156, followed by temperature (F = 3.5) and UV irradiation (F = 3.1) as already shown in Figure 2. The effect of pH was not found significant in TOC reduction (F = 0.02). The effect of Fenton concentration in percentage TOC reduction was more pronounced (almost twice) than temperature and UV irradiation variables and six times more than pH variable. Figure 2

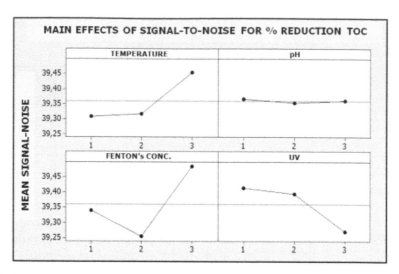

FIGURE 2: Effect of independent variable on the percentage reduction of TOC in terms of S/N ratio (the larger is the best) by photo-Fenton process according to L9 Taguchi matrix.

reveals the variability of the process parameters by average answers graph according to the signal-to-noise ratio (S/R). The best pool of conditions (temperature 35°C, 144 mL of H_2O_2 and 153 mL of Fe^{2+}) was found for percentage TOC reduction in lignin solution after AOP. Figures 1 and 2 clearly reveal the influence of UV irradiation on the percentage reduction of TOC at lowest and highest value. At the highest level of UV irradiation, a remarkable decrease in TOC was observed.

11.2.3 CHANGES IN TOTAL PHENOLS CONCENTRATION

Table 5 shows the average reduction percentage of phenol compounds obtained from double process of experimental conditions in lignin residue treatment, according to L9 orthogonal design. The maximum reduction in TP level from lignin solution was 98.65% in experiment 7 (Table 5). Figure 3 shows the main effect of four process variables (temperature, pH, Fenton reagent concentration and UV radiation) on percentage reduction of TP concentration in lignin solution.

TABLE 5: Average value of total phenols percentage reduction in specific factors and levels to L_9 Taguchi experiments after lignin residue treatment using photo-Fenton process.

Exp.	Temperature (°C)	pH	Fenton (mg L^{-1})	UV (W)	Average of total phenols (%)
1	1	1	1	1	71.10
2	1	2	2	2	84.78
3	1	3	3	3	95.98
4	2	1	2	3	94.08
5	2	2	3	1	96.49
6	2	3	1	2	97.02
7	3	1	3	2	98.65
8	3	2	1	3	97.30
9	3	3	2	1	66.99

TABLE 6: Percentage reduction values of total phenol concentration lignin solution in all 9 experiments carried out as per L_9 Taguchi matrix.

Exp.	TOC range (%)		Average (%)	Si^{2}* (%)
	1° Data set	2° Data set		
1	70.60	71.60	71.10	0.50
2	85.30	84.25	84.78	0.55
3	96.00	95.95	95.98	0.00
4	94.15	94.00	94.08	0.01
5	96.17	96.80	96.49	0.20
6	97.03	97.00	97.02	1.46
7	97.79	99.50	98.65	2.06
8	97.80	96.80	97.30	0.50
9	65.98	68.00	66.99	2.04

*Si^{2} = variance..

The experiments 6 and 8 have shown similar phenolics degradation behavior, with around 97% of average reduction of TP. The experimental conditions which were studied for both experiments demonstrated the temperature range was 30-35°C, pH between 3,0-3,5, low level of Fenton reagent for both experiments (96 mL of H_2O_2 and 87 mL of Fe^2) and irradiation presence in UV (16 - 28W) were not so effective as compared with the experiment 7. Experiment 7 (temperature 35°C, pH 2.5, Fenton reagent H_2O_2 144 ml + Fe^{2+} 153 mL and UV irradiation 16W, proved maximum reduction in TP (98.65%). This combination has shown the best degradation result in the phenol compounds from lignin residue. Table 6 presents the percentage range, the average percentage and the variance (Si^2) in TP concentration in lignin solution after AOP in all 9 experiments performed according to L9 matrix. It is clear from Table 6 that experiment 7 and 9 showed the higher values of standard deviation, 2.06 and 2.04 respectively. Table 7 shows the ANOVA of each process variables involved in lignin residue treatment using photo-Fenton process, according to L9 Taguchi experimental matrix.

It can be observed that all the factors were significant for the studied process parameters in L9 Taguchi matrix. The most significant factor was UV irradiation with F value of 404.00 followed by Fenton reagent concentration (F = 218.80) and temperature (F = 203.1). It can also be observed from the Table 6, that the experiments where UV irradiation were used (experiments 2, 3, 4, 6, 7 and 8) showed an average 95% of removal of phenolic compounds from the lignin solution. Fenton reagent concentration was the second most influence factor (Table 7). Fenton reagent at

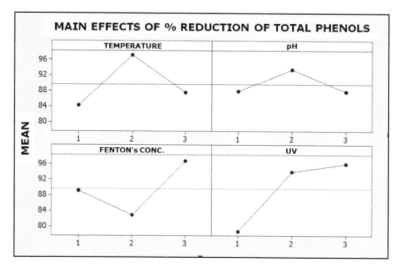

FIGURE 3: Main effects of independent variables (temperature, pH, Fenton reagent concentration and UV irradiation) on total phenols percentage reduction from lignin solution by photo-Fenton process, according to L_9 Taguchi matrix design of experiment.

TABLE 7: Analysis of variance (ANOVA) obtained from average values of total phenols percentage reduction from L_9 orthogonal design to lignin residue treatment using photo-Fenton process.

Factors	Square sum	Degree of freedom	Average square sum	F	p
Temperature	542.35	2	271.2	203.10	0.0000
pH	127.92	2	64.0	47.90	0.0000
Fenton concentration	584.42	2	292.2	218.80	0.0000
Residual	1078.91	2	539.5	404.00	0.0000
UV	12.02	9	1.30	-	-

the concentration (96 mL of H_2O_2 and 87 mL of Fe^{2+}) with UV presence and pH range (3.0-3.5) has shown about 97% removal of total phenolics from lignin solution. Fenton reagent concentration at an intermediate level (120 mL of H_2O_2 and 120 mL of Fe^{2+}) in the UV presence, showed average 89% of removal of phenolic compounds. In experiment 9, where the Fenton reagent were used at an intermediate level, the average of phenolic

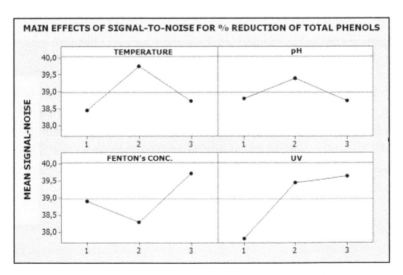

FIGURE 4: Main effects of process variables (temperature, pH, Fenton reagent concentration and UV radiation) on total phenols percentage reduction from lignin solution by photo-Fenton process, according to L_9 Taguchi matrix design of experiment. S/N ratio (the larger is the best) to total phenol percentage reduction.

compounds removal was about 67% probably due to the UV absence in the reaction medium. UV promotes the larger production of •OH radicals which facilitates the large organic material degradation. Figure 4 shows the influence of process variables showing the S/N ratio. Clearly, the greater total phenol reduction percentage (larger signal) is the better. It is clear from the figure that temperature at level 2 (30°C) was more effective than level 1 (25°C) and level 3 (35°C). The effect of pH on percentage TOC reduction was not significant. Concentration of Fenton at level 3 showed remarkable percentage reduction in total phenolics concentration. Similarly, UV ranges at level 3 (28W) showed pronounced effect on total phenolics concentration. Summarizing the effect of these parameters on total phenolics concentration according to L9 Taguchi matrix design of experiment, it can be concluded that the larger S/N ratio (the larger is the best) had the strong influence on total phenolics percentage reduction.

11.3 MATERIALS AND METHODS

11.3.1 RAW MATERIAL

Sugarcane bagasse was acquired from Santa Fé S/A Plant, Nova Europa/SP city. The sugarcane bagasse was used in this experimental as it was obtained from sugarcane processing mill. The processing unit milled the bagasse before sending to laboratory. However, it was not further milled. Thus, sugarcane bagasse was non-uniform in size. It was sun-dried prior to acid hydrolysis in order to remove the extra water. Prior to use in experiment, the moisture content of sugarcane bagasse was analyzed. The total moisture level in bagasse was 10%. This humidity was taken into consideration in acid hydrolysis experiment.

11.3.2 DILUTE ACID HYDROLYSIS

The dilute acid hydrolysis of SB was carried out in a hydrolysis reactor of capacity of 350 L located at the Engineering School of Lorena (EEL)-USP, Lorena, Brazil. For the hydrolysis of the SB, H_2SO_4 (98% of purity) was used as catalyst in a ratio of 100 mg of acid/g of dry matter, during 20 min at 121°C, using a ratio of 1/10 between the bagasse mass and the volume of acid solution. The hydrolysate obtained was maintained at 4°C. The recovered solid residue so called cellulignin was washed with running tap water until the neutralized pH, sun-dried to remove the moisture and was subsequently used for delignification experiments.

11.3.3 DILUTE SODIUM HYDROXIDE PRETREATMENT OF CELLULIGNIN

Cellulignin recovered after dilute acid hydrolysis was submitted to alkaline hydrolysis to recover the lignin solution. For the alkaline hydrolysis, sodium hydroxide (1M) was mixed with cellulignin in the ratio of 1:10. Alkaline hydrolysis was performed in rotatory reactor of capacity 50 L. The reaction was performed at 121°C for 10 min. After cooling the reaction mixture, the liquid fraction (soluble lignin) was separated and subsequently was used for advanced oxidation process (AOP).

11.3.4 COLOR DETERMINATION

Color of lignin solution after alkaline hydrolysis (native) and after AOP was determined by UV-visible spectrophotometer (Bel Photonics, Italy), by using interpolation of an analytical calibration curve as a function of absorbance measurement at 400 nm of wavelength using platinum-cobalt standards [30].

11.3.5 TURBIDITY DETERMINATION

Turbidity measurement of lignin solution after alkaline hydrolysis (native) and after AOP was determined by using a turbidity meter with a precision of 2% (Tecnopon TB 1000). For the calibration of the equipment, standards of 0.1 to 1000 Turbidity Unit of standard solution (Formazin at different dilutions) were used.

11.3.6 CHEMICAL OXYGEN DEMAND (COD) AND BIOCHEMICAL OXYGEN DEMAND (BOD) ANALYSIS

All analytical determinations were performed according to the Standard Methods of Examination of Water and Wastewater [40]. The COD determination is based on the oxidation of organic matter by reduction of potassium dichromate in an acidic medium containing catalyst at elevated temperature and subsequent reading absorbance at a wavelength of 620 nm [41].

In this procedure, the sample was heated for 2 hrs with potassium dichromate in a closed system. During this reaction, oxidizing organic compounds reduce with the conversion of dichromate ion into chromic ion green. The reagents used also contain mercury and silver ions. Silver is a catalyst, and mercury is used to control interference chloride. To determine the accuracy of the method, potassium bi-phthalate (850 mg/L) was used as a standard [42].

In the digestion flasks, 2.0 mL sample, 0.5 mL of digesting solution and 2.5 mL of the catalyst solution were mixed. The mixture was heated at 150°C for 2h. After the cooling to ambient temperature, absorbance was read at 620 nm. The O_2 concentration of the sample was obtained by interpolating the data obtained from the calibration curve performed with a standard solution of potassium bi-phthalate. Standards were prepared of standard values of COD (20-1065). The BOD test is based on the amount of oxygen needed by microorganisms to for the degradation of organic compounds. Organic compounds that were not biodegradable were considered to have BOD value as "zero" [30].

11.3.7 TOTAL ORGANIC CARBON (TOC) DETERMINATION

TOC in lignin solution after alkaline hydrolysis (native) and after AOP was detected by TOC analyzer (TOC-VCPH, Shimadzu, Japan).

11.3.8 TOTAL PHENOLICS (TP) ESTIMATION

TP amount in lignin solution (native) and after AOP was analyzed by the colorimetric method using ferricyanide and antipyrine without extraction (0 - 5.0 mg/L) [30].

11.3.9 ADVANCED OXIDATIVE PROCESS (FE^{2+}/H$_2$O$_2$ AND UV IRRADIATION)

AOP of alkaline hydrolysate of sugarcane bagasse (lignin solution) was performed in a tubular photochemical reactor (GPJ-463/1, Germetec S/A, Brasil) with a volume of approximately 1L at the irradiation of low pressure mercury lamp (GPH-463T5L, Germetec S/A, Brasil), emitting UV radiation at 254 nm with potentials of 16 and 28W, protected by a quartz

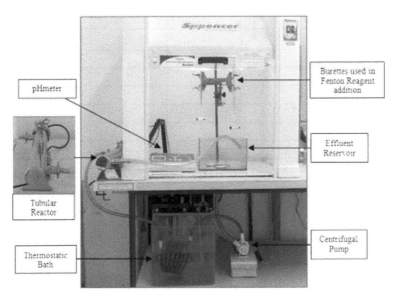

FIGURE 5: Lay-out of the experimental procedure to the treatment step of lignin residue using AOPs.

TABLE 8: Process variables and their ranges selected for the AOP with photo-Fenton process applied to lignin solution using L_9 Taguchi matrix.

Factors	Level 1	Level 2	Level 3
Temperature (°C)	25	30	35
pH	2.5	3.0	3.5
Fenton Reagent * $(m_{H2O2} \ g + m_{Fe+2} L^{-1})$	96 mL H_2O_2 +	120 mL H_2O_2 +	144 mL H_2O_2 +
	87 mL sol Fe^{2+}	120 mL Fe^{2+}	153 mL Fe^{2+}
Ultra-violet irradiation	None	16 W	28 W

*Both H_2O_2 and Iron II were used in solutions with concentrations of $[H_2O_2]$, 30% m/m and $[Fe^{2+}]$, 0.82 mol/L. These values are the proportional to the treated effluent volume of 3 L.

TABLE 9: Experimental plans considering L_9 orthogonal Taguchi design divided in 3 levels (low represented by 1, intermediary represented by 2 and high represented by 3), to the lignin residue treatment using photo-Fenton.

Exp.	Temperature (°C)	pH	Fenton reagent	UV
1	1	1	1	1
2	1	2	2	2
3	1	3	3	3
4	2	1	2	3
5	2	2	3	1
6	2	3	1	2
7	3	1	3	2
8	3	2	1	3
9	3	3	2	1

tube. The reactor configuration as shown in Figure 5, was used for AOP using 3L of lignin solution. The photochemical treatment was carried out in a batch mode. For the Fenton reaction, the volume of the reagents (H_2O_2, 30% and $FeSO_4$.7H2O, 0.18 mol/L) were added. H_2O_2 was added first as drops followed by the addition of $FeSO_4$.7H$_2$O in the lignin solution. AOP was continued up to 2 hrs and the samples were withdrawn after every 10 min. These aliquots were previously adjusted to a pH between 7.0 and 8.0

TABLE 10: Cost-economic analysis of some advanced oxidation processes used for various kinds of wastes.

Type of process	Type of waste	Estimated cost for the process used	Reference
UV/H_2O_2 and UV/TiO_2	Synthetic water	Electricity costs were assumed fixed over the 15 yr at £0.09 per KW h and hydrogen peroxide fixed at £270 m^{-3}	[43]
H_2O_2/UV and Fe^{2+}/H_2O_2	Polyester and acetate fiber dyeing effluent	0.23 ($m-3) and 1.26 ($m^{-3})	[44]
Fe^{2+}/H_2O_2	4-Chlorophenol and Olive oil	Fe^{2+} 0.2 and H_2O_2 10 (kg m^{-3}) Fe^{2+} 0.7 and H_2O_2 18 (kg m^{-3})	[45]
Photocatalysis	Phenol waste water	43.36 ($/L) for all the process	[46]
Photocatalysis	Sewage effluent	15.10 ($/L) for all the process	[47]
Fenton reagent	Phenol waste water	14.28 ($/1000 gallon)	[48]
Photo-Fenton	Tannery waste water	64.13 (US$ m^{-3})	[49]
Photo-Fenton	Pesticides	1.1-1.9 € m^{-3}	[50]

for precipitation of iron salts and then subjected to analysis of COD, color, turbidity and TOC.

11.3.10 EXPERIMENTAL DESIGN

To optimize the effect of process variables (pH, temperature, H_2O_2 and Fenton's Fe^{2+} concentrations and UV radiation potential), a factorial statistical design (L9 Taguchi orthogonal array design of experiments) was performed. Percentage reduction in total phenolics concentration and TOC reduction after AOP were the responsive variables (Tables 8 and 9).

11.3.11 ECONOMIC FEASIBILITY OF THE PROCESS

Two things are needed for any technology to be suitable for use in the industry—the technical feasibility and the economic feasibility. Table 10 shows some examples of cost economic analysis of AOP process employed for the treatment of various kinds of wastes generated.

11.4 CONCLUSIONS

Second generation ethanol production based bio-refineries generate high amount of lignin solution due to the alkali mediated delignification pre-treatment step. Lignin solution is extremely of high recalcitrance with high BOD/COD ratio of 0.25. An ordinary biological pretreatment is not enough to degrade the phenolics amount and consequently bring down the TOC of lignin solution up to satisfactory levels. Therefore, the involvement of AOP is highly recommended to facilitate the phenolics degradation and consequently bring down the TOC levels in lignin solution, recovered from biomass pretreatment. This study clearly demonstrates that AOP mediated by photo-Fenton reaction is highly efficient for the removal of organic material in conjunction with bring down the levels of TOC from sugarcane bagasse alkaline hydrolysate (lignin solution). Taguchi L9 experimental matrix proved that UV irradiation was the most significant factor with F value of 404.00, which significantly removed 98.65% TP and 95.73% of TOC reduction. The results have proved the efficiency of photochemical technology (AOP) in the treatment of residues generated during the alkali mediated delignification pretreatment of SB.

REFERENCES

1. Goldemberg J, Coelho S, Guardabassi P: The sustainability of ethanol production from sugarcane. Ener Pol 2008, 36:2086-2097.
2. Ojeda K, Ávila O, Suárez J, Kafarov V: Evaluation of technological alternatives for process integration of sugarcane bagasse for sustainable biofuels. Chem Eng Res Design 2011, 89:270-279.
3. Chandel AK, Chan EC, Rudravaram R, Narasu ML, Rao LV, Ravindra P: Economics and environmental impact of bioethanol production technologies: an appraisal. Biotechnol Mol Biol Rev 2007, 2:014-032.
4. Cardona CA, Sánchez ÓJ: Fuel ethanol production: process design trends and integration opportunities. Bioresour Technol 2010, 98:2415-2457.
5. Chandel AK, Silva SS, Carvalho W, Singh OV: Sugarcane bagasse and leaves: foreseeable biomass of biofuel and bio-products. J Chem Technol Biotechnol 2012, 87:11-20.
6. Rezende CA, Lima MA, Maziero P, Azevedo ER, Garcia W, Polikarpov I: Chemical and morphological characterization of sugarcane bagasse submitted to a delignification process for enhanced enzymatic digestibility. Biotechnol Biofuels 2011, 4:54.

7. Giese EC, Pierozzi M, Dussan KJ, Chandel AK, Silva SS: Enzymatic saccharification of acid-alkali pretreated sugarcane bagasse using commercial enzyme preparations. J Chem Technol Biotechnol 2012, 88:1266-1272.

8. Canilha L, Chandel AK, Milessi TSS, Antunes FAF, Freitas WLC, Felipe MGA, Silvio SS: Bioconversion of sugarcane biomass into ethanol: an overview about composition, pretreatment methods, detoxification of hydrolysates, enzymatic saccharification, and ethanol fermentation. J Biomed Biotechnol 2012, 1:15.

9. Mussatto SI, Dragone G, Guimarães PMR, Silva JPA, Carneiro LM, Roberto IC, Vicente A, Domingues L, Teixeira JA: Technological trends, global market, and challenges of bio-ethanol production. Biotechnol Adv 2010, 28:817-830.

10. Zhu JY, Pan XJ: Woody biomass pretreatment for cellulosic ethanol production: Technology and energy consumption evaluation. Bioresour Technol 2010, 101:4992-5002.

11. Wang L, Chen H: Increased fermentability of enzymatically hydrolyzed steam-exploded corn stover for butanol production by removal of fermentation inhibitors.

12. Proc Biochem 2011, 46:604-607.

13. Canilha L, Santos VTO, Rocha GJM, Silva JBA, Giulietti M, Silva SS, Felipe MGA, Ferraz A, Milagres AMF, Carvalho W: A study on the pretreatment of a sugarcane bagasse sample with dilute sulfuric acid. J Ind Microbiol Biotechnol 2011, 38:1467-1475.

14. Chandel AK, Silva SS, Singh OV: Detoxification of lignocellulose hydrolysates: Biochemical and metabolic engineering towards white biotechnology. Bio Ener Res 2013, 6:388-401.

15. Rodrigues GD, Silva LHM, Silva MCH: Alternativas verdes para o preparo de amostras e determinação de poluentes fenólicos em água. Química Nova 2010, 33:1370-1378.

16. Uğurlu M, Gürses A, Doğar Ç, Yalçın M: The removal of lignin and phenol from paper Mill effluents by electrocoagulation. J Environ Manag 2008, 87:420-428.

17. Uğurlu M, Karaoğlu MH: TiO2 supported on sepiolite: preparation, structural and thermal characterization and catalytic behaviour in photocatalytic treatment of phenol and lignin from olive mill wastewater. Chem Eng J 2011, 166:859-867.

18. Cansado IPP, Mourão PAM, Falcão AI, Ribeiro Carrott MML, Carrott PJM: The influence of the activated carbon post-treatment on the phenolic compounds removal. Fuel Proc Technol 2012, 103:64-70.

19. Gonçalves MR, Costa JC, Marques IP, Alves MM: Strategies for lipids and phenolics degradation in the anaerobic treatment of olive mill wastewater. Water Res 2012, 46:1684-1692.

20. Amendola D, De Faveri DM, Egües I, Serrano L, Labidi J, Spigno G: Autohydrolysis and organosolv process for recovery of hemicelluloses, phenolic compounds and lignin from grape stalks. Bioresour Technol 2012, 107:267-274.

21. Babuponnusamia A, Muthukumar K: Advanced oxidation of phenol: A comparison between Fenton, electro-Fenton, sono-electro-Fenton and photo-electro-Fenton processes. Chem Eng J 2012, 183:1-9.

22. Oller I, Malato S, Sánchez-Pére JA: Combination of advanced oxidation processes and biological treatments for wastewater decontamination - a review. Sci Total Environ 2011, 409:4141-4166.

23. Fatta-Kassinos D, Vasquez MI, Kümmerer K: Transformation products of pharmaceuticals in surface waters and wastewater formed during photolysis and advanced oxidation process - Degradation, elucidation of by products and assessment of their biological potency. Chemosphere 2011, 85:693-709.

24. Lamsal R, Walsh ME, Gagnon GA: Comparison of advanced oxidation processes for the removal of natural organic matter. Water Res 2011, 45:3263-3269.

25. Wols BA, Hofman-Caris CHM: Review of photochemical reaction constants of organic micropollutants required for UV advanced oxidation processes in water. Water Res 2012, 46:2815-2827.

26. Sharma VK, Triantis TM, Antoniou MG, He X, Pelaez M, Han C, Song W, O'Shea KE, de La Cruz AA, Kaloudis T, Hiskia A, Dionysiou DD: Destruction of microcystins by conventional and advanced oxidation processes: A review. Sep Purif Technol 2012, 91:3-17.

27. Tobaldi DM, Tucci A, Camera-Roda G, Baldi DG, Esposito L: Photocatalytic activity for exposed building materials. J European Ceramic Soc 2008, 28:2645-2652.

28. Michalska K, Miazek K, Krzystek L, Ledakowicz S: Influence of pretreatment with Fenton's reagent on biogas production and methane yield from lignocellulosic biomass. Bioresour Technol 2012, 119:72-78.

29. Cortez S, Teixeira P, Oliveira R, Mota M: Evaluation of Fenton and ozone-based advanced oxidation process as mature landfill leachate pre-treatments. J Environ Manag 2011, 92:749-755.

30. Chu L, Wang J, Dong J, Liu H, Sun X: Treatment of coking wastewater by an advanced Fenton oxidation process using iron powder and hydrogen peroxide. Chemosphere 2012, 86:409-414.

31. APHA, American Public Health Association: Standard Methods for Examination of Water and Wastewater. 21st edition. Washington, DC: (APHA, AWWA); 2005:2001-3710.

32. Lucas MS, Peres JA, Amor C, Prieto-Rodríguez L, Maldonado MI, Malato S: Tertiary treatment of pulp mill wastewater by solar photo-Fenton. J Hazard Mat 2012, 225–226:173-181.

33. Ma Y, Chang C, Chiang Y, Sung H, Chao AC: Photocatalytic degradation of lignin using Pt/TiO2 as the catalyst. Chemosphere 2008, 71:998-1004.

34. Ninomiya K, Takamatsu H, Onishi A, Takahashi K, Shimizu N: Sonocatalytic-Fenton reaction for enhanced OH radical generation and its application to lignin degradation. Ultrasonics Sonochem 2013, 20:1092-1097.

35. Makhotkina OA, Preis SV, Parkhomchuk EV: Water delignification by advanced oxidation processes: Homogeneous and heterogeneous Fenton and H2O2 photo-assisted reactions. Appl Catal B: Environmental 2008, 84:821-826.

36. Pupo Nogueira RF, Trovó AG, Silva MRA, Villa RD: Fundamentos e aplicações ambientais dos processos Fenton e foto-Fenton. Química Nova 2007, 30:400-408.

37. Manenti DR, Gomes LFS, Borba FH, Módenes NA, Espinoza-Quiñones FR, Palácio SM: Otimização do processo foto-Fenton utilizando irradiação artificial na degradação do efluente têxtil sintético. Engevista 2010, 12:22-32.

38. Hermosilla D, Merayo N, Ordóñez R, Blanco A: Optimization of conventional Fenton and ultraviolet-assisted oxidation processes for the treatment of reverse osmosis retentate from a paper Mill. Waste Manag 2012, 32:1236-1243.

39. Samet Y, Hmani E, Abdelhédi R: Fenton and solar photo-Fenton processes for the removal of chlorpyrifos insecticide in wastewater. Water 2012, 38:537-542.
40. Bentivenga G, Bonini C, D'Auria M, De Bona A: Degradation of steam-exploded lignin from beech by using Fenton's reagent. Biomass Bioener 2003, 24:233-238.
41. Salazar RFS, Peixoto ALC, Izário Filho HJ: Avaliação da metodologia 5220 D. Closed reflux, colorimetric method para determinação da demanda química de oxigênio (DQO) em efluentes lácteo. Analytica 2010, 44:55-61.
42. Companhia de Tecnologia de Saneamento Ambiental: Variáveis de Qualidade das Águas. 2013. Disponível em: [http://www.cetesb.sp.gov.br/Agua/rios/variaveis.asp#dbo] Accessed on April, 2013
43. CONAMA - Conselho Nacional do Meio Ambiente: CONAMA - Conselho Nacional do Meio Ambiente. [http://www.mma.gov.br/conama] Accessed on April, 2013
44. Autin O, Romelot C, Rust L, Hart J, Jarvis P, MacAdam J, Parsons SA, Jefferson B: Evaluation of a UV-light emitting diodes unit for the removal of micropollutants in water for low energy advanced oxidation processes. Chemosphere 2013, 92:745-751.
45. Azbar N, Yonar T, Kestioglu K: Comparison of various advanced oxidation processes and chemical treatment methods for COD and color removal from a polyester and acetate fiber dyeing effluent. Chemosphere 2004, 55:35-43.
46. Canizares P, Paz R, Sáez C, Rodrigo MA: Costs of the electrochemical oxidation of wastewaters: A comparison with ozonation and Fenton oxidation processes. J Environ Manag 2009, 90:410-420.
47. Chen YC, Smirniotis P: Enhancement of photocatalytic degradation of phenol and chloro-phenols by ultrasound. Ind Eng Chem Res 2002, 41:5958-5965.
48. Chong MN, Sharma AK, Burn S, Saint CP: Feasibility study on the application of advanced oxidation technologies for decentralized wastewater treatment. J Cleaner Prod 2012, 35:230-238.
49. Mahamuni NN, Adewuyi YG: Advanced oxidation processes (AOPs) involving ultrasound for waste water treatment: A review with emphasis on cost estimation. Ultrason Sonochem 2010, 17:990-1003.
50. Módenes AN, Espinoza-Quiñones FR, Borba FH, Manenti DR: Performance evaluation of an integrated photo-Fenton – Electrocoagulation process applied to pollutant removal from tannery effluent in batch system. Chem Eng J 2021, 197:1-9.
51. Pérez JAS, Sánchez IMR, Carra I, Reina AC, López JLC, Malato S: Economic evaluation of a combined photo-Fenton/MBR process using pesticides as model pollutant. Factors affecting costs. J Haz Mat 2013, 244–245:195-203.

PART IV

OPTIONS FOR THE FUTURE

CHAPTER 12

COMPARATIVE ANALYSIS OF ELECTRICITY COGENERATION SCENARIOS IN SUGARCANE PRODUCTION BY LCA

JOÃO PAULO MACEDO GUERRA, JOSÉ ROBERTO COLETA JR., LUIZA CARVALHO MARTINS ARRUDA, GIL ANDERI SILVA, AND LUIZ KULAY

12.1 INTRODUCTION

Approximately four decades ago, discussions began throughout the world regarding the incorporation of renewable assets into the energy matrix, whether due to the economic instability of crude oil and an increasing rate of fossil resource depletion or the resulting environmental effects. Thus, it was starting a new environmental culture, in which industrial processes were carried out without concerning about the environmental impact (Gil et al. 2013). At present, the use of cleaner energy sources has become a crucial issue for modern society (Gonzáles-García et al. 2012; Luo et al. 2008).

The current energy supply problems in association with the climate change awareness have been motivating the academy to look for renewable fuels and energy sources and put more efforts on studying its process energy efficiency and environmental impacts. Life Cycle Assessment (LCA) has been used more abundantly to assess the environmental impacts of process modifications (Gaudreault et al. 2010). In the past few years, several authors have addressed the thermodynamic analysis of renewable fuels, and others have addressed its environmental performance. Researchers, like Cavalett et al. (2013), have addressed the issue by doing a comparative LCA of ethanol versus gasoline in Brazil using different Life Cycle Impact Assessment (LCIA) methods, concluding that the use of different LCIA methods leads to different conclusions. In the particular case of ethanol and cogeneration plants, Ometto et al. (2009) has performed an LCA of fuel ethanol from sugarcane in Brazil, showing that the fuel ethanol life cycle contributes negatively to all impact potentials analyzed, but in terms of energy consumption, it consumes less energy than its own production largely because of the electricity cogeneration system. Nguyen and Gheewala (2008) have done an LCA of fuel ethanol from cane molasses in Thailand, concluding that the LCA helps to identify the key areas in the ethanol production to improve environmental performance. Renouf et al. (2011) generated an attributional LCA for products produced from Australian sugarcane, showing that the sugarcane products are influenced by some factors, like the nature of a cane-processing system, the variability in sugarcane growing, etc.

Within the Brazilian scenario, the energy industry has one of the highest participation rates of renewable energy on the planet. In 2010, the contributions of hydroelectricity, biomass (using wood and charcoal), biofuels, and wind energy totaled 45.3 % of the country's energy production (MME 2011). Within this context, participation from the sugarcane industry is important because Brazil is the world leader in sugarcane production. Between 2011 and 2012, the national sugarcane production surpassed 530 million tons of raw agricultural materials (MAPA Ministry of Agriculture and Supply 2011). Sugarcane molasses are consumed in the production of ethanol (for use in vehicles) and sugar. However, the remainder of the sugarcane tissue (graminea) is composed of bagasse.

Until recently, the bagasse was discarded in an indiscriminate manner that resulted in significant environmental impacts. Technologies have been developed to change this situation by reusing the bagasse to produce steam and electricity for the sugar and alcohol production processes, making these plants self-sufficient with respect to energy. Thus, sugarcane is potentially important. In addition, cogeneration has become one of the most efficient technologies for the rational use of primary fuel to produce electricity and heat (Bocci et al. 2009; Tina and Passarello 2011).

The prospect of exporting electricity to the national grid also during the idle periods of distilleries and annex plants motivates the Brazilian sugar/alcohol sector to invest in these systems. This study aims to provide basis and decision-making components to this initiative by evaluating actions (within technical and environmental standards) for improving the performance of electrical energy cogeneration units by burning sugarcane bagasse.

12.2 ANALYSIS METHODS

The evaluation of technical performance consisted of the following stages: (1) proposing improvement actions for a cogeneration system based on the Rankine cycle—traditionally employed by Brazilian sugarcane mills. These actions, as well as the thermodynamic operating conditions of the cogeneration system, will be organized as scenarios in order to facilitate the analysis of performance to be made later. (2) It also includes analyzing the potential energy efficiency results, (3) analyzing the environmental performance of the product system from an LCA perspective, and finally (4) being able to get indications that point to the best arrangement for the production of electricity through cogeneration from sugarcane bagasse, both within the fields of energy efficiency and environmental performance. More information about the product system is described in Fig. 1 and discussed further ahead in the "Goal and scope definition" section.

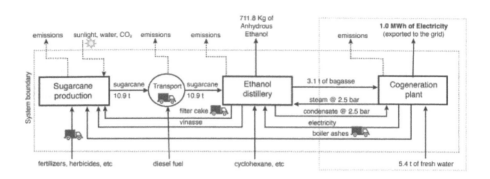

FIGURE 1: Description of the product system under study.

12.2.1 IMPROVEMENT ACTION PROPOSITIONS: THE CASE SCENARIOS

In order to achieve the objective mentioned above, two operating conditions for the cogeneration system were analyzed, including the properties of the steam as it leaves the boiler—at 67 bar and 480 °C or 100 bar and 520 °C, respectively. These operating pressures are realistic and were established from the current trend of the Brazilian sugarcane sector (Dias et al. 2011). For each condition, different energy recovery alternatives were formulated. They were expressed as conventional, reheating, regenerative, and composite (reheating+regeneration) cycles. The arrangement established between the thermodynamic operating conditions, and the alternatives for energy recovery, provided the formulation of eight study scenarios. These scenarios are detailed in Table 1.

The theoretical basis for proposing such arrangements for cogeneration systems came from the experience of efficiency improvement in thermoelectric plants by using the same technology described by Moran & Shapiro (2008).

In the conventional cycle depicted in Fig. 2, superheated steam from the steam generator expands through the turbine until the condenser pressure is reached. The cycle is complete when the turbine exhaust steam condenses and returns to the boiler with the process condensate. In the reheating cycle depicted in Fig. 3, the steam does not expand to reach the

TABLE 1: Electricity production scenarios for the cogeneration systems based on different steam properties and energy recovery conditions.

Case scenario	Description
Standard (SS)	Conventional vapor power cycle at 67 bar and 480 °C
I	Reheating vapor power cycle at 67 bar and 480 °C
II	Regenerative vapor power cycle at 67 bar and 480 °C
III	Reheat–regenerative vapor power cycle at 67 bar and 480 °C
IV	Conventional vapor power cycle at 100 bar and 520 °C
V	Reheating vapor power cycle at 100 bar and 520 °C
VI	Regenerative vapor power cycle at 100 bar and 520 °C
VII	Reheat–regenerative vapor power cycle at 100 bar and 520 °C

condenser pressure in a single stage. Instead, the steam expands through a first-stage turbine and is reheated in the steam generator. The flow after reheating is reinjected into the turbine in the second stage. The regenerative cycle assumes intermediate withdrawals of steam along the turbine. These steam flows are used to heat the boiler feedwater, which occurs through indirect contact in heat exchangers as showed in Fig. 4. Finally, the composite cycle consists of reheating and regeneration in a single cycle as depicted in Fig. 5.

The potential gains achieved by the improvement actions were analyzed by comparing the overall energy efficiency results and the amount of electricity delivery to the power grid in each scenario with those of the standard scenario (SS).

12.2.2 THERMODYNAMIC ANALYSIS

The models developed to represent the conventional cycle along with the steam reheating, regeneration, and composite cycles are represented in Figs. 2, 3, 4, and 5. For all case scenarios, dry saturated steam was extracted from the turbine at a pressure of 2.5 bar for using in the ethanol production process and for reaching the set deaeration point (110 °C) before returning to the boiler. The turbine exhaust steam flows into a condenser unit at a pressure of 0.1 bar. The liquid fraction passes through a deaerator prior to reintroduction into the boiler at the end of the cycle.

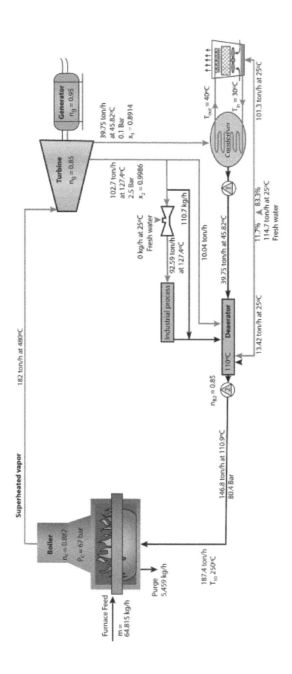

FIGURE 2: Conventional steam power cycle for a cogeneration plant.

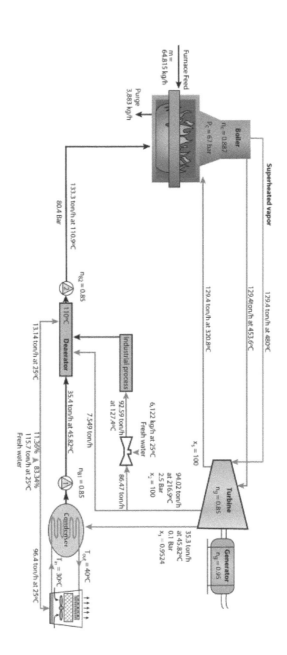

FIGURE 3: Reheating steam power cycle for a cogeneration plant.

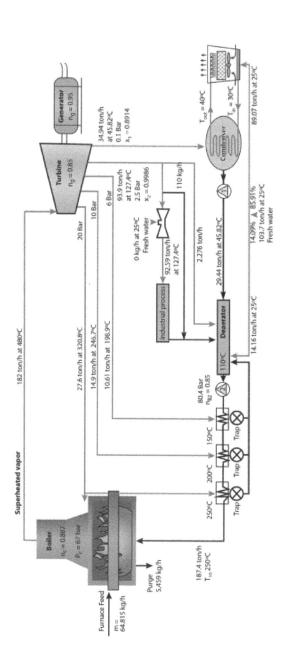

FIGURE 4: Regenerative steam power cycle of a cogeneration plant.

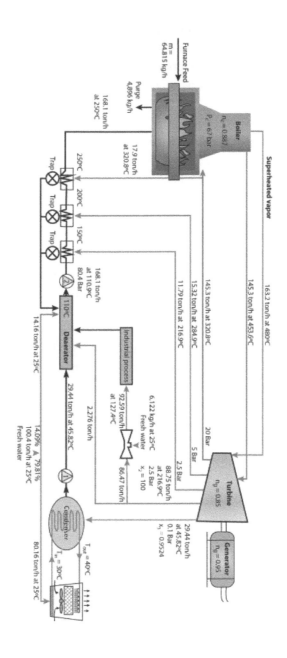

FIGURE 5: The reheat–regenerative steam power cycle for a cogeneration plant.

In the reheating cycle (Fig. 3), superheated steam is extracted from the turbine at an optimum pressure of 20 to 25 bar for both work conditions in this study. This variation was determined from parametric analysis by applying the graphic inspection method. The turbine exhaust steam followed the same trajectories as that in the conventional cycle. In the regenerative cycle (Fig. 4), the number of heat exchangers was also determined by parametric analysis, and it was realized that the use of more than three units in series to heat the feedwater has slightly increased the efficiency of the regenerative cycle. In light of this finding, it was decided to keep an arrangement with three heat exchangers units.

The main thermodynamic indicators analyzed were the energy efficiency, the net power output, and net power exported to the grid. The energy efficiency of a cogeneration system was defined as the ratio between the useful energy—i.e., the thermal energy used for ethanol production plus the total electricity obtained by the cogeneration—and the total amount of energy entered in the boiler. The net power output consists of the difference between the total energy produced in the cogeneration unit and the power consumed within the Rankine cycle by the pumps. The net power exported is the power delivered to grid, and can be determined by the difference between the net power output, and the power required for ethanol production.

The case scenarios were modeled using the Engineering Equation Solver (EES) software. This computational resource is useful to provide thermodynamically robust solutions for energy systems. For this purpose, the EES solves a set of algebraic equations, and it is particularly useful for design problems in which the effects of one or more parameters need to be determined. In this paper, EES was applied to make an analysis varying just one parameter at a time in a suitable range in order to find the optimum result. The equations used in the models were based on the principles of conservation of mass and energy applied for each step of the Rankine cycles.

12.2.3 ENVIRONMENTAL ASSESSMENT

To evaluate the environmental performance, an LCA was used to focus on the cogeneration plant "from cradle-to-gate". The goal and scope of this evaluation was to study the production and delivery of electricity to the grid. Under these conditions, the product system considered sugarcane in the agricultural production stage, its transport to the distillery, its use for anhydrous ethanol production, and its use for the cogeneration of electrical energy. For the product system model, it was based on the agricultural practice regularly developed in the state of São Paulo where the primary data were collected. The average agricultural productivity of Sao Paulo reaches 90 t/ha, and it is responsible for 61 % of total Brazilian sugarcane production (CGEE 2008). In addition, the reuse of industrial residues, such as filter cakes, vinasse, and boiler ashes, and burning of the growing area in preparation for harvest were considered (Sousa and Macedo 2010). Details and conditions associated with this methodological step will be provided below.

12.3 GOAL AND SCOPE DEFINITION

The calculation employed in the simulations for all scenarios was based on anhydrous ethanol production in an autonomous sugarcane distillery with a crushing capacity of 2.0 Mt of sugarcane per season (180 days).

It corresponds to a sugarcane bagasse input of 64,815 kg/h in the boiler furnace. It was considered that 100 % of the bagasse is sent to the boiler and converted into thermal energy. It means that bagasse does not cross the system boundary to be sold or disposed.

The thermal energy consumption (400 kg of saturated vapor at 2.5 bar) and electricity (30 kWh/t of processed sugarcane) required for ethanol production are not affected in the previously described scenarios. The excess electrical energy generated by the cogeneration unit is exported to the electricity grid.

The LCA study was conducted based on the theoretical registration described in ISO 14040 (2006a) and ISO 14044 (2006b). Thus, based on the objective definition, the initiative proposes to conduct an environmental analysis of actions that improve the performance of cogeneration units by burning sugarcane bagasse. Regarding the scope definition, the following conditions were established.

- *Functional unit*—to the delivery of 1.0 MWh of electricity to the power grid using a cogeneration system under the conditions mentioned above.
- *Product system*—includes the sugarcane agricultural production stages, transport, the industrial production of ethanol, and the electricity cogeneration. The diagram of the product system appears in Fig. 1. The ethanol production rate was kept constant for all the scenarios, once the processing alternatives under study have been simulated only at the cogeneration plant, always admitting the same sugarcane crushing rate.
- *Data source*—secondary data served to model the product systems, with the exception of the SS, for which primary data were collected and were referred to as the equipment's performance data.
- *Data quality*—the Temporal Coverage consisted of the 2-year period of 2009 and 2010. The geographical coverage comprised the state of Sao Paulo. Technical coverage considered the processes and technical features previously described.
- *Allocation*—the environmental load from the agricultural step and industrial process of ethanol production are allocated between ethanol and sugarcane bagasse. Thereby, it was taken into account an allocation criterion based on energy content, which was expressed in terms of the lower heating value (LHV) of the components.
- *Types of impact and methodology of LCIA*—to obtain an environmental performance profile with a wide spectrum as generated by grouping analytical indicators, the method ReCiPe Midpoint (H) version 1.08 was selected (Goedkoop et al. 2013).

All of the impact categories from ReCiPe were considered, except for ionization radiation, marine eutrophication and marine ecotoxicity, ozone and metal depletion and urban land occupation. The features of the object of study and the character analysis supported this decision.

Contributions in terms of ionization radiation were not taken into account due to the low expressivity of modal nuclear emission in the national energy matrix. Developments in the form of eutrophication and ecotoxicity on marine biota are outside the scope of application of the study, as well as urban land occupation.

Depletions of ozone and metals were dismissed after the product system model has analyzed its significance in terms of the potential contributors to both environmental impacts. In addition, infrastructure issues, as well as capital goods, were disregarded after the simulations carried out have showed discrete influence on the environmental impacts.

As a premise of the study, it was established that the resource consumption and emissions, as well as the operating conditions of the unit processes of "sugarcane production" and "sugarcane crushing"—which is part of the subsystem "ethanol distillery"—are the same for all scenarios under evaluation. Therefore, the environmental load amounted by the bagasse into the cogeneration plant was either the same for any of the simulations. This assumption was held in order to evaluate the effects in terms of energy efficiency and under the environment of producing increasing electricity in a cogeneration unit using the same amount of natural resources.

12.4 RESULTS

12.4.1 THERMODYNAMIC ANALYSIS

Figures 6 and 7 provide the overall energy efficiency results and the relative gains in terms of the net power exported to the grid for each case scenario (measured relative to the SS). Based on the results, it is possible to see that the energy efficiency improves as the pressure at which the vapor leaves the boiler increases. In addition, the energy efficiency improves during the reheat–regenerative steam power cycle relative to the other cycles. For scenario VII, which had the greatest overall energy efficiency, the net power output reached 152 kWh/t of sugarcane and the net power exported reached 122 kWh/t of sugarcane. This result corresponds to an increase of 33 % relative to the performance of the SS.

Figures 8 and 9 show that the makeup water consumption in the cooling tower was reduced by 32 % from SS to scenario VII. On the other hand, the makeup water supplied to the boiler was increased by 58 % relative to the initial SS. Comparing the best case scenario to the SS, it is possible to notice an overall reduction of 21 % in the water consumption for the Rankine cycle. The introduction of a reheating and regeneration independently to the conventional Rankine cycle at 67 bar caused a reduction of 5 and 12.1 % in terms of cooling water makeup, respectively. Figure 8 also shows that the combination of reheating and regeneration applied to the Rankine cycle (case scenario III) caused a reduction of 21 % of the cooling water makeup.

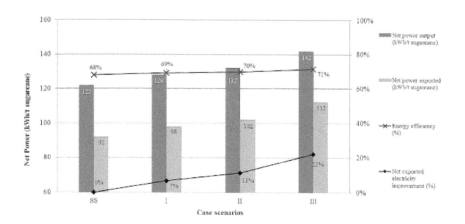

FIGURE 6: Thermodynamic results of the case scenarios at 67 bar in relation to the SS.

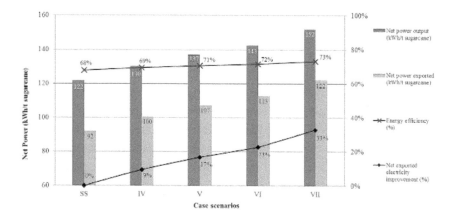

FIGURE 7: Thermodynamic results of the case scenarios at 100 bar in relation to the SS.

The case scenarios at 100 bar showed the same trend. The more energy is lost in the condenser, the more cooling water flow through the condenser is required. The cooling water comes from the evaporative cooling tower. That is the reason of the better the energy efficiency of the case scenario, the less energy wasted in the cooling tower. As a result, the overall fresh-water consumption decreases.

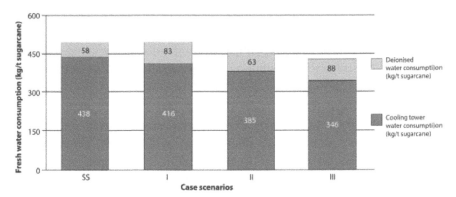

FIGURE 8: Freshwater consumption results of the case scenarios at 67 bar in relation to the SS.

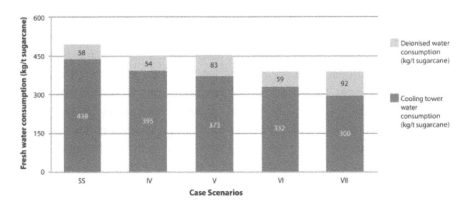

FIGURE 9: Fresh water consumption results of the case scenarios at 100 bar in relation to the SS.

12.4.2 ENVIRONMENTAL ASSESSMENT

The environmental performance of the product system under analysis is indicated in Tables 2 and 3. Table 2 describes, as a Life Cycle Inventory (LCI), the main consumption of resources—both natural, as obtained from the TechnoSphere— and emissions—to air, water, and soil—for the eight operation scenarios of the Rankine cycle under study. Table 3 displays the environmental performance profile results generated by the LCIA for the eight case scenarios using the method ReCiPe Midpoint (H)—version 1.08. Only a broad analysis performed both in the levels of LCI and LCIA enables a clear, comprehensive, and accurate picture of the consequences caused by the electricity production via cogeneration and its delivery to the grid.

From a generalist point of view, this approach revealed unanimously for all the scenarios under study that the agricultural stage—related to the unit process "sugarcane production"—provides more negative effects to the environment than any other elements of the product system. It is followed in descending order of importance for "soil amendment", "transportation", "ethanol distillery", and the "cogeneration plant". In the cogeneration plant, the less significant contributions from all of the product system were recorded. This is justified because the unit presents a productive arrangement that comprises closed loops of thermal and electrical energy and a partial recycle of water.

It is possible to observe from the obtained results that the improving energy performance of the system from the adoption of alternative technologies and the increased vapor pressure output is accompanied by reduced environmental impacts for all evaluated categories. Besides, an analysis performed just under the environmental approach points out that the improvements remain strictly equal in relation to the SS for all the impact categories considered by the study, if they are expressed as percentage contributions. This conclusion can be observed in Fig. 10.

It is important to emphasize that the results presented in Table 3 are strongly dependent of the allocation criteria. As mentioned in the "Goal and scope definition" section, the only allocation of environmental loads performed in this study occurred between the ethanol and bagasse. It was carried out in terms of energy content, expressed as LHV. According to this approach, the bagasse shifts about 50 % of the total environmental load associated to the agricultural process and the industrial stage of sugarcane crushing to the cogeneration plant.

From a more specific perspective, it is observed that 56 % of the contributions for the category of climate change (CC) are regarding to emissions of dinitrogen oxide (N_2O). About 98 % of it occurs in the "sugarcane production" unit process because of activities of burning straw, nitrogen fertilizer oxidation, vinasse and filter cake disposal, as well as waste biomass

TABLE 2: Main environmental impacts associated to the delivery of 1.0 MWh of electricity to the power grid, obtained from cogeneration

Environmental loads	Unit	SS	I	II	III	IV	V	VI	VII
INPUTS									
Inputs from nature									
Gas, natural, and in ground	m³	1.12E+01	1.01E+01	1.05E+01	9.17E+00	1.02E+01	9.11E+00	9.60E+00	8.42E+00
Occupation, arable, and non-irrigated	m²a	7.99E+02	7.19E+02	7.50E+02	6.55E+02	7.32E+02	6.52E+02	6.86E+02	6.02E+02
Oil, crude, and in ground	t	3.04E−02	2.73E−02	2.85E−02	2.49E−02	2.78E−02	2.48E−02	2.61E−02	2.29E−02
Transformation, from arable, nonirrigated	m²	4.23E+00	3.81E+00	3.97E+00	3.47E+00	3.87E+00	3.45E+00	3.63E+00	3.18E+00
Transformation, from pasture and meadow, extensive	m²	1.75E+01	1.57E+01	1.64E+01	1.43E+01	1.60E+01	1.43E+01	1.50E+01	1.32E+01
Transformation, from shrub land, sclerophyllous	m²	2.21E−01	1.99E−01	2.08E−01	1.82E−01	2.03E−01	1.81E−01	1.90E−01	1.67E−01
Transformation, to arable, nonirrigated	m²	2.20E+01	1.98E+01	2.06E+01	1.80E+01	2.01E+01	1.79E+01	1.88E+01	1.65E+01
Water, at the surface	m³	1.29E+01	1.14E+01	1.22E+01	1.03E+01	1.17E+01	9.99E+00	1.09E+01	9.25E+00

TABLE 2: CONTINUED.

Inputs from TechnoSphere									
Agricultural machinery, general	kg	7.28E−01	6.55E−01	6.84E−01	5.97E−01	6.65E−01	5.93E−01	6.24E−01	5.48E−01
Ammonium nitrate phosphate, as N	kg	7.56E−01	6.80E−01	7.10E−01	6.20E−01	6.91E−01	6.15E−01	6.48E−01	5.69E−01
Diesel	kg	1.75E+01	1.58E+01	1.64E+01	1.44E+01	1.60E+01	1.42E+01	1.50E+01	1.32E+01
Carbofuran	kg	3.83E−02	3.45E−02	3.60E−02	3.14E−02	3.50E−02	3.12E−02	3.28E−02	2.88E−02
Diuron	kg	8.90E−03	8.01E−03	8.36E−03	7.30E−03	8.13E−03	7.24E−03	7.63E−03	6.70E−03
Fipronil	kg	3.65E−03	3.29E−03	3.43E−03	2.99E−03	3.34E−03	2.97E−03	3.13E−03	2.75E−03
Glyphosate	kg	2.08E−02	1.87E−02	1.95E−02	1.71E−02	1.90E−02	1.69E−02	1.78E−02	1.57E−02
Growth regulators	kg	7.71E−03	6.94E−03	7.24E−03	6.32E−03	7.05E−03	6.28E−03	6.61E−03	5.81E−03
Harvester	kg	2.46E−01	2.21E−01	2.31E−01	2.02E−01	2.25E−01	2.00E−01	2.11E−01	1.85E−01
Lime, hydrated and packed	kg	4.55E+00	4.10E+00	4.27E+00	3.73E+00	4.16E+00	3.70E+00	3.90E+00	3.43E+00
Potassium chloride, as K_2O	kg	6.09E+00	5.48E+00	5.72E+00	4.99E+00	5.57E+00	4.96E+00	5.22E+00	4.59E+00
Single superphosphate, as P_2O_5	kg	2.74E+00	2.47E+00	2.57E+00	2.25E+00	2.50E+00	2.23E+00	2.35E+00	2.06E+00
Urea, as N	kg	5.02E+00	4.52E+00	4.71E+00	4.12E+00	4.59E+00	4.09E+00	4.30E+00	3.78E+00
Tractor	kg	5.12E−01	4.61E−01	4.81E−01	4.20E−01	4.68E−01	4.17E−01	4.39E−01	3.86E−01
Vinasse, from sugarcane	m³	7.15E+00	6.44E+00	6.71E+00	5.86E+00	6.54E+00	5.82E+00	6.13E+00	5.38E+00

TABLE 2: CONTINUED.

OUTPUTS									
Emissions to air									
NH^3	kg	2.19E+00	1.97E+00	2.05E+00	1.79E+00	2.00E+00	1.78E+00	1.88E+00	1.65E+00
Cd	kg	1.61E−04	1.45E−04	1.51E−04	1.32E−04	1.48E−04	1.31E−04	1.38E−04	1.21E−04
Carbon dioxide, biogenic	kg	1.32E+03	1.18E+03	1.23E+03	1.08E+03	1.20E+03	1.07E+03	1.13E+03	9.90E+02
CO_2, fossil	kg	1.50E+02	1.35E+02	1.41E+02	1.23E+02	1.37E+02	1.22E+02	1.29E+02	1.13E+02
CO_2, land transformation	kg	2.94E+02	2.65E+02	2.76E+02	2.41E+02	2.69E+02	2.40E+02	2.52E+02	2.21E+02
CO, biogenic	kg	2.20E+01	1.98E+01	2.06E+01	1.80E+01	2.01E+01	1.79E+01	1.89E+01	1.66E+01
CO, fossil	kg	8.30E+01	7.47E+01	7.79E+01	6.81E+01	7.60E+01	6.77E+01	7.12E+01	6.25E+01
Cu	kg	2.62E−02	2.36E−02	2.46E−02	2.15E−02	2.40E−02	2.13E−02	2.25E−02	1.97E−02
CH_4	kg	3.54E−01	3.19E−01	3.32E−01	2.90E−01	3.24E−01	2.89E−01	3.04E−01	2.67E−01
CH_4, biogenic	kg	6.59E−01	5.93E−01	6.19E−01	5.40E−01	6.03E−01	5.37E−01	5.66E−01	4.96E−01
CH_4, fossil	kg	2.13E+00	1.92E+00	2.00E+00	1.75E+00	1.95E+00	1.74E+00	1.83E+00	1.61E+00
N_2O	kg	2.18E+00	1.96E+00	2.05E+00	1.79E+00	2.00E+00	1.78E+00	1.87E+00	1.64E+00
NMVOC, unspecified origin	kg	4.61E+01	4.15E+01	4.33E+01	3.78E+01	4.22E+01	3.76E+01	3.95E+01	3.47E+01
Particulates, <10 um	kg	2.83E+00	2.55E+00	2.66E+00	2.32E+00	2.59E+00	2.31E+00	2.43E+00	2.13E+00
Particulates, <2.5 um	kg	8.25E+01	7.43E+01	7.75E+01	6.76E+01	7.55E+01	6.73E+01	7.08E+01	6.21E+01
SO_2	kg	1.59E+01	1.43E+01	1.49E+01	1.30E+01	1.46E+01	1.30E+01	1.36E+01	1.20E+01
Zn	kg	1.55E−02	1.39E−02	1.45E−02	1.27E−02	1.42E−02	1.26E−02	1.33E−02	1.17E−02

TABLE 2: CONTINUED.

Emissions to water									
Fipronil	g	5.47E−02	4.92E−02	5.14E−02	4.48E−02	5.01E−02	4.46E−02	4.69E−02	4.12E−02
PO_4^{3-}	g	8.67E+01	7.80E+01	8.15E+01	7.11E+01	7.94E+01	7.06E+01	7.45E+01	6.54E+01
P	g	1.09E+00	9.84E−01	1.03E+00	8.96E−01	1.00E+00	8.91E−01	9.38E−01	8.23E−01
Emissions to soil									
Cd	g	1.10E−01	9.94E−02	1.04E−01	9.05E−02	1.01E−01	9.00E−02	9.47E−02	8.32E−02
Carbofuran	g	3.77E+01	3.40E+01	3.54E+01	3.09E+01	3.45E+01	3.08E+01	3.24E+01	2.84E+01
Diuron	g	8.77E+00	7.89E+00	8.23E+00	7.19E+00	8.02E+00	7.15E+00	7.52E+00	6.60E+00
Fipronil	g	3.59E+00	3.23E+00	3.37E+00	2.94E+00	3.29E+00	2.93E+00	3.08E+00	2.71E+00

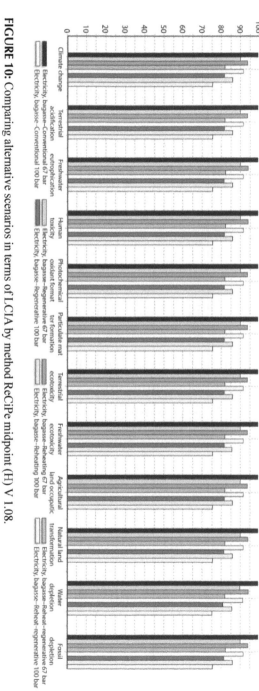

FIGURE 10: Comparing alternative scenarios in terms of LCIA by method ReCiPe midpoint (H) V 1.08.

TABLE 3: Environmental performance profiles for scenarios (delivery of 1.0 MWh of electricity to the grid).

Impact category	Unit	Case scenarios							
		SS	I	II	III	IV	V	VI	VII
CC	kg CO_2 eq	1.17E+03	1.05E+03	1.10E+03	9.60E+02	1.07E+03	9.55E+02	1.00E+03	8.82E+02
TA	kg SO_2 eq	2.24E+01	2.01E+01	2.10E+01	1.83E+01	2.05E+01	1.82E+01	1.92E+01	1.69E+01
FEu	kg P eq	2.54E-02	2.29E-02	2.39E-02	2.09E-02	2.33E-02	2.07E-02	2.19E-02	1.92E-02
HT	kg 1,4-DB eq	4.71E+01	4.23E+01	4.42E+01	3.86E+01	4.31E+01	3.83E+01	4.04E+01	3.54E+01
POF	kg NMVOC	5.47E+01	4.92E+01	5.13E+01	4.48E+01	5.00E+01	4.46E+01	4.69E+01	4.12E+01
PMF	kg PM_{10} eq	8.97E+01	8.07E+01	8.42E+01	7.35E+01	8.21E+01	7.31E+01	7.69E+01	6.75E+01
TEc	kg 1,4-DB eq	3.43E+00	3.08E+00	3.22E+00	2.81E+00	3.13E+00	2.79E+00	2.94E+00	2.58E+00
FEc	kg 1,4-DB eq	1.41E+00	1.27E+00	1.32E+00	1.16E+00	1.29E+00	1.15E+00	1.21E+00	1.06E+00
ALO	m²a	8.03E+02	7.22E+02	7.53E+02	6.58E+02	7.35E+02	6.54E+02	6.89E+02	6.04E+02
NLT	m²	5.21E-02	4.68E-02	4.89E-02	4.26E-02	4.76E-02	4.24E-02	4.46E-02	3.92E-02
WD	m³	1.43E+01	1.26E+01	1.35E+01	1.14E+01	1.29E+01	1.11E+01	1.21E+01	1.03E+01
FD	kg oil eq	2.38E-04	2.15E-04	2.24E-04	1.95E-04	2.18E-04	1.94E-04	2.05E-04	1.80E-04

left on the field, and diesel combustion from agricultural machinery. A parcel of 25.1 % of the impact for the same category can be attributed to carbon dioxide (CO_2) emitted because of land transformation. Considering the model of expansion of sugarcane for the state of Sao Paulo proposed by Macedo et al (2008), the progress of this cultivation in the region will occur in 80 % on pasture areas, in 19 % on other annual crops, and in 1 % on Brazilian savannah along the period 2009–2010. CO_2 emissions derived from land transformation were thus attributed to suppression of vegetation cover from these areas. It should also highlight that 12.8 % of CO_2 emissions are from fossil origin, as a consequence of application of urea and lime to soil in order to adjust the cultivation conditions, in the transport of raw materials, and an intermediate process such as farm machinery. Finally, biogenic methane (CH_4) emitted from straw burning represents 1.3 % of CC impacts.

The most significant contributions as the terrestrial acidification (TA) are associated with emissions of sulfur oxides (SO_x), 71.1 %, and ammonia (NH_3), 23.9 %. Diesel combustion from cultivation, harvesting, and transportation emits 95 % of the total amount of SOx. Redox (reduction–oxidation) reactions that occur with nitrogenous fertilizers—urea, monoamonium nitrate phosphate, and diamonium nitrate phosphate (MAP and DAP)—which remain in the soil and that are exposed to anaerobic atmosphere represent about 99 % of the NH_3 emitted in the "sugarcane production".

Freshwater eutrophication (FEu) displays a defined contribution profile. Phosphate (PO_4^{3-}) releases in water represent 95.5 % of the environmental burden associated with the category. This contribution is mainly due to losses that occur during production of single superphosphate (SSP)—4.42 kg/t—a phosphate fertilizer regularly used for sugarcane cultivation at the Sao Paulo state.

Human toxicity (HT) occurred from various contributions, all of them associated with the sugarcane cultivation. The stand out is, in this case, the emission of cadmium (Cd) and zinc (Zn) in the air (respectively, 12.4 and 16.7 % of contributions) Cd to soil (22.3 %), and phosphorous to water (21.6 %). Diesel combustion for cultivation and harvesting represents more than 94 % of the metal releases in the air. Losses of Cd to soil can be attributed to phosphate fertilizer application. The presence of P in the water originates from the production of growth regulators (67.1 %) and glyphosate (20.5 %) for plague control in the sugarcane cultivation. Glyphosate is an organophosphorus compound obtained from phosphorus tri- and pentachloride—PCl3 and PCl5. Its production contributes with water losses of 10.7 kg P/t (Green 1987). Growth regulators are employed as pesticides to the sugarcane cultivation in Brazil even if this class of compounds does not actually kill any organisms. As in glyphosate, the growth regulators are also obtained from phosphorus chlorides (0.186 kg/ kg), and its production imposes the losses of about 95 kg P/t (Lucas and Vall 1999).

The burning of sugarcane (straw and leaves) to interrupt the process of saccharification or even biomass for land clearing between successive crops is still a recurrent practice in the state of Sao Paulo. Both processes occur under low—or even virtually nonexistent—excess oxygen. In these conditions, nonmethane volatile organic compounds—NMVOC—and fossil carbon monoxide (CO) are produced by incomplete combustion. These environmental loads represent respectively 84.3 and 6.9 % of the contribution for photochemical oxidant formation (POF).

The same activities, in association with incomplete diesel combustion in obsolete agricultural machinery, justify releases of both SOx and particulate matter with specific diameters $\phi < 2.5$ μm and $\phi < 10$ μm into the air. These effects represent respectively 3.6, 92 and 3.2 % of the contribution in terms of particulate matter formation (PMF).

Emission of particles with $\phi < 2.5$ μm occur predominantly at "sugarcane production" (99.4 %). Moreover, atmospheric releases of PM $\phi < 10$ μm arises mainly from burning of bagasse in cogeneration (65.8 %). The remaining 34.2 % proceed from burning of straw in the field. Data presented in Sousa and Macedo (2010) indicate average emissions of 2.77 kg SO_x per ton of sugarcane due to the diesel combustion and, once again, to the burning of straw.

The emission to soil of carbofuran (with a contribution of 55.3 %) fipronil (8.1 %), and diuron (7.5 %), active ingredients used for plague control represent, accounts for the main agents in terms of terrestrial ecotoxicity (TEc). To these, the emission to the air of cooper (Cu) from diesel combustion can still be added, which represents 25.1 % of the total amount attributed to the same impact category. Freshwater ecotoxicity

(FEc) follows a similar profile. To this case, losses of carbofuran to soil contribute 40.1 % of the total impact. It is followed by releases of fipronil (14.8 %) and diuron (7.2 %), both to soil. The emission of P to water and Cu in air—in which contributions are 8.5 and 8.1 %—may also be highlighted as a significant contribution. As mentioned before, for the case of HT, contributions associated to P emission in water come from the production of growth regulators (67.2 %) and glyphosate (20.5 %).

For agricultural land occupation (ALO), 99.6 % of the total contribution calculated for each of the scenarios corresponds to occupation of arable land without irrigation. This result is consistent with the characteristics of extensive cultivation of sugarcane practiced in the state of Sao Paulo for which that classification of land use is quite usual (Sousa and Macedo 2010).

The results in terms of natural land transformation (NLT) can be divided into two categories: indirect and direct effects. Indirect effects represent 75.4 % of total contributions for the category and refer mainly to crude oil circulation (4.2 %) and natural gas (1.3 %), which occur by pipelines, installed onshore; flood areas to form dams in hydroelectric power plants (3.4 %); and onshore drilling wells for crude oil exploration (82.6 %).

Contributions to NLT category regarding hydropower are based on data provided by the National Petroleum, Natural Gas and Biofuels Agency (ANP) in which the contribution of this mode to the national energy matrix corresponded to 81.4 % (ANP 2012). Land transformation due to the drilling of oil wells is because Brazil imported in 2011 about 16 % of crude oil from countries like Nigeria, Algeria, Iraq, Saudi Arabia, and Libya, where the extraction of this natural resource occurs on land (ANP 2011). The direct effects are due to the aforementioned expansion of cultivation of cane sugar over the Brazilian savannah.

Water depletion (WD) is concentrated in 90.4 % of the water consumption pumped mainly from rivers for the production of tap water in order to generate steam in the cogeneration system. To this total, consumption—not greater than 1.5 %—relative to irrigation water to crops of sugarcane can also be added.

Finally, the contributions of the product system for the purpose of fossil depletion (FD) may also be considered negligible. These focus mainly on the consumption of crude oil (87.8 %), from which, diesel to drive farm machinery and transport vehicles and natural gas (about 2.17 %) originates from it—for preparation of synthetic nitrogen fertilizers like urea, MAP, and DAP.

12.5 CONCLUSIONS

Reheating and regeneration concepts were found to be considerably effective in improving the energy efficiency of cogeneration systems by burning sugarcane bagasse. The results of this study confirmed the expectation that the sugar/alcohol sector can increase the amount of electricity delivered to the national grid using the same amount of natural resources and as a result increase revenue.

Alternatives for improving the Rankine cycle also reduced environmental impacts based on the standard analysis of all examined environmental impacts. The study also verified that any improvement in the sustainable use of natural resources is related to the introduction of more efficient practices for cogeneration systems.

The analysis performed by LCA revealed that the main environmental impacts associated to the delivery of 1.0 MWh of electricity to the national grid from cogeneration of sugarcane bagasse appeared along the agricultural production stages. The inclusion of the environmental variable in the evaluation of alternative technologies for improving energy efficiency of the same system brought a different perception for analysis. Through this approach, it was possible to note that the implementation of effective process optimization, even if it was restricted to only one stage of the life cycle, can result in systemic reductions of the negative effects on the environment provided by an anthropic action. For the case in specific, this finding does not exempt the pursuit of technologies, practices, and more sustainable behaviors on the part of decision makers that comprise the sugarcane sector in Brazil. Prior to this, the conclusions signal that the exercise of Life Cycle Thinking should be stimulated within the industry in order to deliver business value.

REFERENCES

1. ANP—National Petroleum, Natural Gas and Biofuels Agency (2011) Brazilian statistical yearbook of oil, natural gas and biofuels. Rio de Janeiro, p 282
2. ANP—National Petroleum, Natural Gas and Biofuels Agency (2012) Brazilian Statistical Yearbook of oil, natural gas and biofuels. Rio de Janeiro. p 280
3. Bocci E, Di Carlo A, Marcelo D (2009) Power plant perspectives for sugarcane mills. Energy 34:689–698
4. Cavalett O, Chagas M, Seabra J, Bonomi A (2013) Comparative LCA of ethanol versus gasoline in Brazil using different LCIA methods. Int J Life Cycle Assess 18:647–658
5. CGEE (Center for Strategic Studies and Management) (2008) Bioethanol sugarcane: energy for sustainable development. BNDES, Rio de Janeiro
6. Dias M, Modesto M, Ensinas A, Nebra S, Filho R, Rossel C (2011) Improving bioethanol production from sugarcane: evaluation of distillation, thermal integration and cogeneration systems. Energy 36(6):3691–3703
7. Gaudreault C, Sanson R, Stuart P (2010) Energy decision making in a pulp and paper mill: selection of LCA system boundary. Int J Life Cycle Assess 15:198–211
8. Gil M, Moya A, Domíngues E (2013) Life cycle assessment of the cogeneration processes in the Cuban sugar industry. J Clean Prod 41:222–231
9. Goedkoop M, Heijungs R, Huijbregts M, De Schryver A, Struijs J (2013) Description of the ReCiPe methodology for life cycle impact assessment. In: ReCiPe main report (revised 13 May 2012). http://www.lcia-recipe.net. Accessed 7 Sept 2012
10. Gonzáles-García S, Iribarren D, Susmozas A, Dufour J, Murphy R (2012) Life cycle assessment of two alternative bioenergy systems involving Salix spp. biomass: bioethanol production and power generation. Appl Energy 95:111–122
11. Green, M. (1987) Energy in pesticide manufacture, distribution and use. In: Helsel ZR (ed.) Energy in plant nutrition and pest control, Vol. 7. Elsevier, Amsterdam, ISBN 0-444-42753-8, pp. 165–177
12. ISO 14040 (2006a) International Organization for Standardization, Environmental management—life cycle assessment—principles and framework. Geneva. p 21
13. ISO 14044 (2006b) International Organization for Standardization, Environmental management—life cycle assessment—requirements and guidelines. Geneva, 52p
14. Lucas S, Vall MP (1999) Pesticides in the European Union. Agriculture, Environment, Rural Development—Facts and Figures. European Communities. http://europa.eu.int/comm/agriculture/envir/report/en/pest_en/report_en.htm
15. Luo L, Voet E, Huppes G (2008) Life cycle assessment and life cycle costing of bioethanol from sugarcane in Brazil. Renew Sustain Energy Rev 13:1613–1619
16. Macedo I, Seabra J, Silva J (2008) Greenhouse gases emissions in the production and use of ethanol from sugarcane in Brazil. The 2005/2006 averages and prediction for 2020. Biomass 7. Energy 32:582–595
17. MAPA (Ministry of Agriculture, Livestock and Supply) (2011) Secretariat and bioenergy production, sugar and ethanol in Brazil. Harvest 2010–2011. Industry statistics, Brasília

18. MME (Ministry of Mines and Energy) (2011) Review Brazilian energy 2010. MME, Brasília
19. Moran J, Shapiro N (2008) Fundamentals of engineering thermodynamics, 6th edn. John Wiley & Sons, New York
20. Nguyen T, Gheewala S (2008) Life cycle assessment of fuel ethanol from cane molasses in Thailand. Int J Life Cycle Assess 13:301–311
21. Ometto A, Hauschild M, Roma W (2009) Lifecycle assessment of fuel ethanol from sugarcane in Brazil. Int J Life Cycle Assess 14:236–247
22. Renouf M, Pagan R, Wegener M (2011) Life cycle assessment of Australian sugarcane products. Int J Life Cycle Assess 16:125–137
23. Sousa E, Macedo I (Org.) (2010) Ethanol and bioelectricity: sugarcane in the future energy mix. Luc Communication Project, São Paulo
24. Tina G, Passarello G (2011) Short-term scheduling of industrial cogeneration systems for annual revenue maximization. Energy 42:46–56

CHAPTER 13

TECHNO-ECONOMIC COMPARISON OF ETHANOL AND ELECTRICITY COPRODUCTION SCHEMES FROM SUGARCANE RESIDUES AT EXISTING SUGAR MILLS IN SOUTHERN AFRICA

ABDUL M. PETERSEN, MATHEW C. ANEKE, AND JOHANN F. GÖRGENS

13.1 BACKGROUND

Sugarcane processing industries in Southern Africa generate bagasse at a yield of 0.30 tons per ton of cane processed [1]. In most sugar mills in Southern Africa, the generated bagasse is mostly burnt to provide heat and electricity for the sugar milling operations [1,2]. South African sugar mills (from crushing to raw sugar production) typically have poor efficiency and the average steam demand is 0.58 tons per ton of sugarcane processed [3] (58% on cane). When such process designs are coupled with low efficiency biomass-to-energy conversion systems, then no surplus bagasse is generated by the sugar mill and therefore no export of electricity occurs [4,5]. If efficient sugar mills that have steam demands below 40% [5,6] are coupled with efficient systems that convert biomass to energy [6], then excess bagasse becomes available. This excess, if combined with other post-harvest residues like sugarcane trash, could provide the feedstock for the production of bio-energetic products in an integrated facility. The costs associated with the utilization of such residues would include the cost of collection and transport, and the investment costs required to upgrade the energy efficiency of existing sugar mills to enable the liberation of surplus bagasse. These costs are significantly lower than the purchasing costs of biomass [7] that hinders the economic viability of 'stand-alone' facilities for biomass conversion to energy [8].

The low efficiency biomass-to-energy systems in older cane milling operations utilized combustion systems that had raised steam to pressures of between 15 and 22 bar [5,9]. Such systems also provided a low cost means of disposing of bagasse [1,9] at a time when exporting electricity was not economically interesting. For that means, combustion with high pressure steam cycles allowed for greater turbine efficiency in the conversion of steam to electricity and thus, pressures of 82 to 85 bar [1,10] would have typically been preferred. At a pressure of 60 bar, it has been shown that a net electricity export of 72 kW (per ton of cane processed per hour) was possible for an efficient sugarcane mill, where a steam demand of 0.4 tons per ton of cane was required [5]. This amount of export electricity could have been increased substantially if the harvesting residues (trash) was also considered [5,7,11]. The electrical efficiencies resulting from biomass power plants utilizing combustion and high pressure steam cycles are reported to be between 23 and 26% on an HHV (Higher Heating Value) basis [12,13], while efficiencies reported for Biomass Integrated Gasification and Combined Cycles system (BIGCC) were at 34 to 40% [14]. The implementation of BIGCC in industry has been limited due to the reportedly high capital investment that is required [12,13,15]. The capital estimates of BIGCC systems in previous techno-economic assessments [12,13,16] however, were based on the estimates in a period where BIGCC technology was still new (1990 to the early 2000s) [17], and thus, capital estimates based on the vendor quotes in this period would have reflected the pioneer plant costs. A capital estimate based on a matured estimate could be significantly lower than the pioneer estimate [18].

As an alternative to the conversion of all of the available lignocellulose residues to electricity, a fraction of the bagasse and post-harvest residues could be used to produce ethanol, with co-generation of electricity. The hemicellulose, which makes up about 20 to 35% [19] of the biomass matrix, can be solubilized by steam explosion or dilute acid hydrolysis and converted to ethanol, while the remaining cellulose-lignin fractions are converted to heat and power [20,21]. This scenario for the coproduction of ethanol and electricity from lignocellulose has been proposed for the South African industry [20], but a detailed process flow sheet and techno-economic investigation of such for existing sugar mills is not available. Of particular interest would be the techno-economic comparison of co-production of ethanol and electricity against a scenario where the residues are used exclusively for electricity generation. Previous studies have compared electricity generation alone with the complete lignocellulose conversion to ethanol (hemicellulose and cellulose) as options for integration with sugar mills [22] and autonomous distilleries [7,11]. The ethanol generation scheme in this study builds on the concept of 'value prior to combustion' that has previously been evaluated as a green-field (stand-alone) scenario [21].

There has been a considerable success in developing microbial strains that efficiently converts pentose-rich hydrolysates to ethanol [23], which is the key area of importance if the proposed technology is to be feasible. Using adapted strains of a the native pentose fermenting yeast *Pichia stipitis*, Kurian et al. [24] converted 82.5% of the hemicellulose sugars in a hydrolysate derived from sorghum bagasse that contained 92 g/l of dissolved sugars, while Nigam [25] converted 80.0% in an acid hydrolysate from wheat straw, containing 80 g/l sugars. The development of robust recombinant strains, such as the *Saccharomyces cerevisiae* TMB400, have resulted in pentose conversions in excess of 85% in toxic environments in simultaneous saccharification and fermentation experiments [26]. More recently, the National Renewable Energy Laboratory (NREL) achieved an ethanol yield of 92% on hemicellulose sugars in a toxic enzymatic hydrolysate that contained a total of about 150 g/l of sugars, using the *Zymomonas mobilis* strain that was genetically engineered by Du Pont [27]. Thus, fermentation technology for converting pentose sugars in hydrolysates to ethanol has been successfully demonstrated on a laboratory scale.

The present study provides a detailed techno-economic comparison of scenarios that entail ethanol coproduction with export electricity, produced either by combustion or BIGCC systems, against those that produce only export electricity using the same systems. For either scenario, the upgrading costs of the existing sugar mill to achieve an energy efficiency of 0.40 ton of steam per ton of cane, is included in the capital investments considered in the economic analysis. The development of process models for the ethanol coproduction scenario will be based on established flow sheets and process performances for lignocellulosic ethanol [28,29], and will also consider various processing options to ensure the most energy efficient and economical flow sheet. The projects are assumed to be in Kwa-Zulu Natal where the sugar cane crushing plants are concentrated. All South African legislations would apply.

Energy efficiency for all of the scenarios will be maximized through pinch point analysis (PPA) for the heat integration of the processing streams [30-32]. This approach will ensure that the energy utilities for ethanol production are kept to a minimum [29,30], consequently maximizing the export electricity while still providing the energy requirements of the (energy efficient) mill [33]. From the process simulations (mass- and energy-balances) for the various scenarios economic evaluations, incorporating capital and operational costs as well as sales prices, will be performed from an economic risk perspective [34-37]. These methods are based on Monte Carlo simulations that are super-imposed on standard methods for process economic methods, in order to account for the risks associated with the fluctuations in economic variables, thereby providing not only the estimates for investment returns, but also the probability of achieving economic success.

13.2 RESULTS AND DISCUSSION

13.2.1 TECHNICAL EVALUATION

Six scenarios for the production of electricity from sugarcane residues, either as the only energy product or with coproduction of ethanol from hemicellulose, were evaluated through process modelling to estimate process energy efficiency and economics. The results of the energy characteristics for the various process alternatives that have been optimized by

TABLE 1: Bio-energetic product yields, utility demands and energy efficiencies (optimized by pinch point analysis) (CD/CHPSC – Ethanol Production with Conventional Distillation with energy generation from pretreatment residues using CHPSC; CD/BIGCC – Ethanol Production with Conventional Distillation with energy generation from pretreatment residues using BIGCC; VD/CHPSC – Ethanol Production with Vacuum Distillation with energy generation from pretreatment residues using CHPSC; VD/BIGCC – Ethanol production with Vacuum Distillation with energy generation from pretreatment residues using BIGCC; CHPSC-EE – Exclusive electricity generation from residues using CHPSC; CHPSC-EE (Dryer) – Exclusive electricity generation from residues using CHPSC where biomass is dried; BIGCC-EE – Exclusive electricity generation from residues using BIGCC)

Scenarios	Ethanol cogeneration				Exclusive electricity		
	CD/CHPSC	CD/BIGCC	VD/CHPSC	VD/BIGCC	CHPSC-EE	CHPSC-EE (Dryer)	BIGCC-EE
Net Outputs							
Bioethanol Production (l/hr)	9601	9577	9599	9575			
Electricity Production (MW)	22.06	33.94	23.42	46.47	51.00	55.85	88.63
Steam Generation and Requirements							
Gross Steam Generation (tons/hr)	204.58	146.43	204.58	146.42	277.22	294.71	191.98
Mill Steam Demand (tons/hr)	120.00	120.00	120.00	120.00	120.00	120.00	120.00
Ethanol Generation Steam Demand (tons/hr)	51.06	51.06	30.25	30.25			
Steam Contingency (tons/hr)	33.53	−24.63	54.34	−3.83	157.22	174.71	71.98
As percentage of Total Steam Demand	19.60	−14.40	36.17	−2.55	131.02	145.59	59.99
Electricity Generation and Requirements							
Gross Electricity Generation (MW)	38.52	187.50	40.76	187.51	65.49	70.50	263.36
Power (MW)	1.17	122.63	1.36	122.63	2.00	2.15	162.24
Ethanol Utilities (MW)	2.81	18.44	3.49	5.91			
Mill Electricity Demand (MW)	12.49	12.49	12.49	12.49	12.49	12.49	12.49
Net Energy Efficiency	25.57%	29.17%	25.98%	32.91%	15.44%	16.86%	26.83%

pinch point analysis are presented in Table 1. Furthermore, the amount of steam generated by the heat and power facility in each scenario, whether this facility forms an exclusive electricity scenario or an energy generation section of an ethanol coproduction scenario, is presented. If the facility utilizes the Combustion with High Pressure Steam Cycles (CHPSC) technology, then the gross steam generation refers to gross amount of steam generated by the biomass-fired boiler. If the heat and power plant utilizes the BIGCC technology then the gross steam generation refers to the steam generated by the heat recovery steam generator (HRSG) that recovers heat from the gas turbine's exhaust. The steam contingency refers to the amount of steam that is reserved once all the demands of the sugar mill and ethanol plants (in the case of ethanol coproduction scenarios) are met, and is essentially an indication of the operating leeway the scenario offers in terms of meeting steam when fluctuations in the plant occur. According to Pellegrini et al. [38], the maximum fluctuation of the steam demand in a sugar mill was measured at 2%.

The ethanol production rate of all the coproduction scenarios averages 9591 l/hr, which would equate to 62 million liters per annum. Given that the total consumption of road transport fuel in South Africa is about 23 billion liters per annum [39], this production rate would represent 0.27% of road transport fuels. This production rate equates to an ethanol yield of 35 liters per ton of cane crushed, where the hemicellulose fraction of the bagasse and 50% of the trash generated is converted to ethanol. With regards to the exclusive electricity generation, the BIGCC-EE and CHPSC-EE (EE - exclusive electricity production using BIGCC and CHPSC respectively) scenarios generated 88.63 MW and 53.43 ± 2.43 MW of electricity (MWe) respectively. Given that the total output of electricity supplied to the national grid is 34 GW [40], then the contribution to the grid would be 0.22% and 0.13% for the BIGCC-EE and CHPSC-EE respectively. The coproduction of ethanol with electricity from sugarcane residues available at sugar mills would reduce the potential electrical export by approximately 54% on average.

With regards to the steam generation and demand, it is seen that the gross generation of steam in the BIGCC-EE process is 32.86% less than the amount of the CHPSC-EE process, primarily because BIGCC technology is meant to maximize electricity generation, rather than steam generation.

The steam generated by the heat and power generation facilities of the ethanol coproduction scenarios are 28.4% lower when compared to their exclusive electricity counterparts. The major implication of this reduction for steam generation was that when BIGCC technology was coupled with ethanol coproduction, the combined steam demand of the sugar mill and the ethanol generation process exceeded the steam generated. This penalized the electricity generated by the ethanol-BIGCC scenarios, as electricity was needed internally for heating purposes, at a rate of 2.43 MW and 15.63 MW for the VD/BIGCC and CD/BIGCC respectively. Thus, the vacuum distillation scenarios offered a more feasible operating scenario when BIGCC technology was considered, as the lower steam consumption minimized the electricity consumed for heating purposes.

A further comparison of vacuum and conventional distillation shows that the application of vacuum distillation allowed for an extra 1.36 MW of electricity to be available for export (comparing CD/CHPSC and VD/CHPSC). This was because the lower steam demand of the vacuum distillation system on the steam utilities allowed for more steam to expand through the exhaust steam turbine of the condensing extraction steam turbine (CEST). Furthermore, the multi-effect system also relieved the cooling duty of the condenser of the rectifier column, and thus, no further electricity was needed to deliver this cooling duty. So even though there was an additional process electricity requirement for the vacuum pump that actuated the multi-effect distillation, the reduction in utility requirements exceeded the requirement, which then resulted in a net positive electricity export.

The energy efficiencies reported in Table 1 were based on the net export of the bio-energetic products, which is the ethanol sold and the electricity exported to the grid after the mill requirements were accounted for. Generally, ethanol scenarios with BIGCC technology had the greatest net export efficiency, followed by that of the BIGCC-EE, which was comparable to the ethanol-CHPSC scenarios, and the lowest being the CHPSC-EE scenario. The reduction in steam consumption in vacuum distillation when CHPSC and BIGCC technology are used for energy generation is shown to improve the export electricity efficiency by 0.41% and 3.74% respectively. The improvement when BIGCC technology was integrated with vacuum distillation is explained by the lower amount of electricity consumed internally for heating purposes. The lowest export energy efficiency was attained by the CHPSC-EE scenario, due to the large amount of exhaust steam still present after the steam demand of the sugar mill was accounted for. The energy contained in this steam is mostly spent to the environment by the surface condenser. If a biomass dryer was used to de-moisture the biomass prior to combustion, as in the case of the CHPSC-EE, then the export efficiency improved by 1.42% because the steam and electricity generation had improved by 6.14% and 7.65% respectively.

In order to assess the benefits of pinch point analysis (PPA), values in Table 1 are compared with the corresponding values in Table 2. Table 2 does not report values for the exclusive electricity scenarios because effective heat integration is implicit in the overall design of these processes. Regarding the CHPSC-EE, the design of Nsaful et al.[41], which is the source model for the CHPSC technology, was already optimized with PPA. As for the BIGCC-EE, PPA confirmed the heat integration strategies that have previously been implemented, such as the cooling of the syngas to heat up air for the gasifier, improving steam generation, and also using the inter-cooler duty of the multistage compressor to improve steam generation [37].

The comparison between Tables 1 and 2 shows that the potential for exporting electricity from the ethanol-BIGCC scenarios are reduced by a margin of 30 to 38% if PPA is not applied, mainly because the increase in steam demand resulted in more electricity consumed for heating purposes. Thus, the primary advantage of PPA is the reduction of steam and electrical utilities, which then resulted in the net export of more electricity, and the general improvement in the export energy efficiencies by 0.98% and 3.99% for ethanol-CHPSC and ethanol-BIGCC scenarios respectively. The effect of PPA on the scenario employing conventional distillation seems more apparent because the vacuum distillation already affected a substantial reduction in utilities by inducing multi-effect distillation.

13.2.2 ECONOMIC RESULTS

The six scenarios for the production of electricity from sugarcane residues, either as the only energy product or with coproduction of ethanol from

TABLE 2: Yields, utility demands and energy efficiencies without pinch point analysis

Scenarios	CD/CHPSC	CD/BIGCC	VD/CHPSC	VD/BIGCC
Net Outputs				
Bioethanol Production (l/hr)	9601	9577	9599	9575
Electricity Production (MW)	18.63	52.38	20.37	52.38
Steam Generation and Requirements				
Gross Steam Generation (tons/hr)	204.58	146.43	204.58	146.42
Mill Steam Demand (tons/hr)	120.00	120.00	120.00	120.00
Ethanol Generation Steam Demand (tons/hr)	68.62	68.62	49.56	49.56
Steam Contingency (tons/hr)	15.97	-42.19	35.02	-23.14
As percentage of Total Steam Demand	8.47	-22.37	20.66	-13.65
Electricity Generation and Requirements				
Gross Electricity Generation (MW)	36.63	187.50	38.99	187.51
Power (MW)	1.01	122.63	1.21	122.63
Ethanol Utilities (MW)	4.50	31.29	4.92	19.61
Mill Electricity Demand (MW)	12.49	12.49	12.49	12.49
Net Energy Efficiency	24.53%	34.70%	25.05%	34.70%

hemicellulose, were compared in terms of the total capital investments (TCI) required (Figure 1), the economic viabilities in terms of internal rates of return (IRR) on the investments (Figure 2), and the financial risk of each investment, based on the Monte Carlo simulation, quantified as the probability of an acceptable return on investment (Figure 3). Figure 1 shows that the highest capital investment was 324.57million US$ for the VD/BIGCC cogeneration scenario, which was also the scenario with the highest energy efficiency. The primary reason for the high capital investment is the costs associated with the integrated BIGCC, as shown

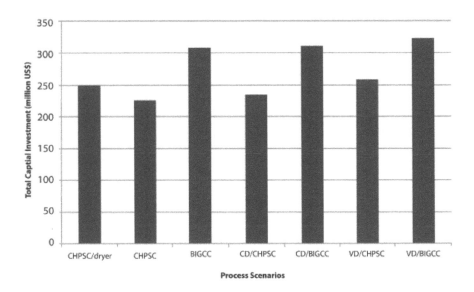

FIGURE 1: Total capital investment for simulated scenarios. (CD/CHPSC – Ethanol Production with Conventional Distillation with energy generation from pretreatment residues using CHPSC; CD/BIGCC – Ethanol production with Conventional Distillation with energy generation from pretreatment residues using BIGCC; VD/CHPSC – Ethanol Production with Vacuum Distillation with energy generation from pretreatment residues using CHPSC; VD/BIGCC – Ethanol production with Vacuum Distillation with energy generation from pretreatment residues using BIGCC; CHPSC-EE – Exclusive electricity generation from residues using CHPSC; CHPSC-EE (Dryer) – Exclusive electricity generation from residues using CHPSC where biomass is dried; BIGCC-EE – Exclusive electricity generation from residues using BIGCC).

by the difference in capital costs between the ethanol scenarios with vacuum distillation that have either the CHPSC or BIGCC technologies integrated as energy islands. The application of the vacuum distillation also demanded higher capital costs, as is shown by the general comparison of

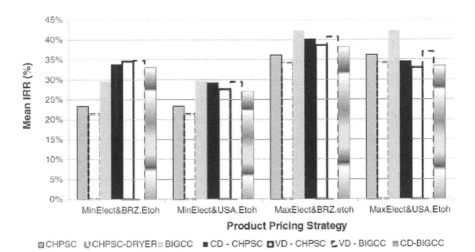

FIGURE 2: Comparison of processes profitability at varies strategies. "Mean IRR" refers to the average of the IRRs determined from each iteration of the Financial Risk Simulation.

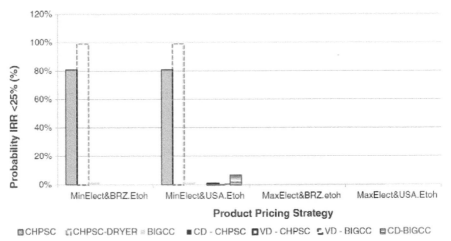

FIGURE 3: Evaluation of financial risk at varies pricing strategies. The value on the vertical axis describes the probability to which the IRRs simulated for each scenario will fall below a standard value of 25% for the IRR.

the vacuum and conventional distillation scenarios. With VD, additional capital charges were also incurred by the larger capacity of the surface condenser and circulation pumps. This was because of the lower steam demand, which resulted in a greater throughput of steam in the equipment mentioned. Thus, the use of VD is shown to increase the capital expenditure by 9.33% when CHPSC technology is used, and 4.33% when the BIGCC technology is used as energy schemes, respectively.

The TCI of BIGCC-EE, which was based on a modern estimate [42], is 28.89% higher than the average TCI of CHPSC-EE. This difference is much less than those attained in previous comparative studies that based the BIGCC capital estimates on pioneer costs. The relative difference between BIGCC and the CHPSC technologies in Dornburg et al. [43] and Bridgwater et al. [12] was about 50% and 77% respectively. The difference in the comparison by Bridgewater et al. [12] had been exceptionally large, since the scale on which the comparison was based was a 20 MWe. Trends shown by Bridgewater et al.[12] and Dornburg et al. [43] have indicated that the difference gets smaller as the scale increases.

With regards to exclusive electricity production, it is shown (Figure 2) that the lowest IRRs were attained by the CHPSC-EE scenario, both inclusive and exclusive of the biomass dehydration prior to combustion. The IRRs obtained for the combustion scenarios did not differ significantly from each other, and on average were 22.38% (±0.98%) and 35.20% (±0.97%) at the minimum and maximum premiums on electricity, respectively. In either case, the higher value was obtained when no dryer was considered. Thus, the energy efficiency gained by employing a biomass dryer was not economically justified, due to the increased capital expenditure. The IRR attained by the BIGCC-EE is 7% and 6% higher than those attained by the CHPSC-EE scenario, when the minimum and maximum premiums on electricity were considered, respectively. In previous studies where the capital estimates of the BIGCC technology were based on pioneer quotes, the profitability of CHPSC technology was generally higher [12,43]. As the capital estimate of the BIGCC technology in this study was based on a modern estimate, it shows that BIGCC-EE became more favorable, which had also been demonstrated by Searcy and Flynn [15].

It was calculated that on average, the ethanol prices projected from Brazilian data were 47% greater than those based on US data, and therefore the overall minimum pricing strategy was the minimum electricity premium for electricity with US-based ethanol prices. At this pricing scenario, the highest profitability of the ethanol scenarios was attained by the VD/BIGCC. The economic feasibility of integrating BIGCC technology as the heat and power system of ethanol coproduction is advantageous over the CHPSC technology as the profitability of the VD/BIGCC was higher than the VD/CHPSC. This result was expected since the BIGCC-EE was more profitable than the CHPSC-EE when the exclusive electricity scenarios were compared due to the larger surplus of electricity. With regards to the implementation of VD, the IRR of the VD/CHPSC scenario was 1.05% lower than that of the CD/CHPSC, whereas the IRR of the VD/BIGCC was 1.48% higher than the CD/BIGCC. Thus, VD was not justified by the additional capital costs when integrated with CHPSC technology, but was more profitable when integrated with BIGCC technology. Due to the high amount of electricity that was needed for heating purposes in the CD/BIGCC scenario, the IRRs was lower than the VD/BIGCC, even though the capital expenditure was less.

When the maximum electricity premium was considered, the IRR of the BIGCC-EE was 1.59% higher than the most profitable ethanol scenario (VD/BIGCC) obtained with the higher (Brazilian) ethanol prices. Under the lower (US) ethanol pricing, the BIGCC-EE was 5.44% higher than the VD/BIGCC scenario (still the most profitable of the ethanol scenarios). Regarding the maximum electrical premium however, the IRRs shown in Figure 2 for exclusive electricity scenarios are well in excess of the standard IRR of 17% that is imposed by the National Energy Regulator of South Africa (NERSA) for independent power producers (IPPs) [44]. In order to diversify the renewable energy contribution of the South African electricity supply (for example, from solar or wind), the prices paid to IPPs are regulated by NERSA to maintain an IRR of 17%, in order to promote equal investment opportunity in the various forms of renewable electricity [44]. Thus, as both the BIGCC-EE and CHPSC-EE are shown to meet this target at the minimum electricity premium, a higher price for electricity generated in the sugar mills would not be allowed.

With regards to assessing the scenarios from a financial risk perspective (Figure 3), the maximum occurrence at which the IRR can be less than 25% (which is known to attract the interests of private investors [36,45]) was a probability of 20% [41]. This qualification of 20% is an extension of a general criterion applied for the maximum probability of the net present value (NPV) being less than zero [41]. At the optimistic pricing scenarios, where high premiums on electricity are considered, all scenarios would be attractive for private investment (Figure 3) when the IRRs are evaluated against the IRR standard of 25%. An evaluation at the minimum pricing scenario showed that all ethanol coproduction scenarios qualified for private investment, since the risks associated with an unfavorable return for private investment were generally less than 1%. The maximum risks were attached to exclusive electricity production involving combustion, as the probabilities of the IRR falling below the standard of 25% were above 80.82% for the CHPSC-EE without drying and 98.96% with drying.

The status of the high risk imposed by the exclusive electricity scenarios to private investment would not improve due to restrictions imposed by NERSA on IPPs in respect of the standard IRR of 17%. However, sensitivity in the ethanol prices could allow for private investment that is virtually risk free when coproduction of ethanol is considered, as shown when ethanol prices are projected from the Brazilian data. Under that circumstance, the risks of an unfeasible return for a private investor for all the ethanol scenarios are acceptable, even when the minimum premium for electricity is considered.

13.2.3 COMPARISON OF THE PRESENT STUDY WITH SIMILAR STUDIES IN PUBLISHED LITERATURE

Since the CHPSC-EE scenario was modelled on the flow sheet of Nsaful et al. [41], the results of this scenario was validated with the technical and economic results of the scenario using combustion and 82 bar steam cycle, in that study. The export efficiency calculated based on the electrical export of 86.02 kW per ton of cane amounts to 12.70%, which is lower than the export efficiency of 15.44% reported for the CHPSC-EE of this study. The export efficiency of this study is higher because the electricity generation was supplemented with sugarcane trash, which improved the amount of electricity available for export. The IRR reported for the process modelled by Nsaful et al. [41] was 29%, when a bagasse and electricity price of 56 US$/dry ton (data from 2010) and 0.248 U$ per kWhr are considered respectively. Under these conditions, the model in this study yielded an IRR of 41.82%. The optimistic outcome arose because the supplementation by trash improved the export electricity to 170.01 kW per ton of cane processed. Furthermore, the incorporation of trash, which costs just 30% of the bagasse price, reduces the average specific cost paid for biomass in the model of this study to 41.53 U$/dry ton (data from 2010).

The BIGCC-EE scenario was compared with the results of Craig and Mann [46], who conducted a techno-economic study of various options for a BIGCC power plant fuelled by wood. Options explored included various gasification scenarios, such as pressurized versus near-atmospheric conditions, and direct versus indirect heating. Since this plant was an autonomous facility, the energy demands of the sugar mill were discarded in order to remodel the BIGCC-EE scenario as an autonomous facility, so that the results could be comparable. The net electrical efficiency of the autonomous BIGCC-EE was 34.2%, which compared well with the value of 37.9% obtained by Craig and Mann [46]. The efficiency of Craig and Mann [46] is expected to be higher because the combined steam cycle of the pilot plant operated at the much higher pressure of 100 bar, which was a more efficient system for the steam cycle than the steam cycle in this study, which operated at a steam pressure of 60 bar. Thus a greater contribution was expected from the steam cycle section in that study [41]. Furthermore, Craig and Mann [46] assumed an efficient design of the gasifier that assured complete conversion of the biomass, whereas this study considered a conservative case where only 90 to 95% of the biomass was converted.

The minimum electricity price (MEP) determined by Craig and Mann [46] was 0.07 US$ per kWhr on a currency base of 1996, which is equivalent to a 2012 price of 0.132 US$ per kWhr. This was attained under the economic constraints of a biomass price of 78.67 US$ per dry ton (US$/d-ton) and an IRR of 10%. With such constrains, the exclusive BIGCC-EE of this study was shown to obtain a MEP of 0.10 US$ per kW/hr. The discrepancy then arises from the maturity of the capital cost estimates, as Craig and Mann [46] used very early stage estimates of BIGCC systems (1990), which was 51.97% more than the modern estimate (2008) used in the BIGCC scenarios of this study.

The results of the CD/CHPSC was compared to the feasibility assessment of an integrated ethanol facility conducted by Macrelli et al. [11], where the cellulose fractions of bagasse and trash residues were considered for second generation ethanol production. The net export efficiency of the integrated component, based on the export of electricity and lignocellulosic ethanol generated, was 35.2%, which is significantly higher than the efficiency of 25.57% determined for the CD/CHPSC. This arose because Macrelli et al. [11] considered the cellulose fraction for ethanol production, which constitutes 35 to 40% of the considered biomass, as opposed to hemicellulose which only constitutes 20 to 24% of the considered biomass (see Table 3). Thus, a greater fraction of the biomass was efficiently used for the production of the energetic product.

The MESP (Minimum Ethanol Selling Price) that was determined for the lignocellulosic ethanol by Macrelli et al. [11] was 0.97 US$ per litre, under the economic constraints of a sugarcane trash price of 26 US$ per dry ton and an IRR of 10%. This MESP included a penalty of 0.12 US$ per litre for the reduction of export electricity when compared to electricity exports of an autonomous first generation facility and an enzyme cost of 0.38 US$ per litre. Thus, the MESP was adjusted to 0.47 US$ per litre by disregarding the penalty and enzyme cost, for a consistent comparison with this study. Under the economic constraints in the study of Macrelli et al. [11], the CD/CHPSC obtained a MESP of 0.43 US$ per litre, which compares well with the adjusted MESP of Macrelli et al.[11]. This comparison also shows that although a process converting cellulose to ethanol is more energy efficient, it is less economically viable due to the major cost associated with the enzymes needed to hydrolyse cellulose.

13.2.4 SENSITIVITY ANALYSIS

Figure 4 and Figure 5 show the economic sensitivity of the economic parameters to certain parameters that were deemed to have an uncertainty in their specification. The sensitivity was carried out with the pricing strategy that considered the minimum selling prices of electricity and ethanol. The coproduction scenarios with the highest profitability were the VD/BIGCC and the CD/CHPSC, which were also the best indications of ad-

TABLE 3: Chemical composition of sugarcane residues.

Component (%)	Bagasse[1]	Trash[2]
Cellulose	41.1	39.8
Hemicelluloses	26.4	28.6
Lignin	21.7	22.5
Ash	4.0	2.4
Extractives	6.8	6.7

[1]Average of measurements for South African bagasse [47,48].
[2]Composition taken Oliveira et al, [49].

vanced and immediate technologies available, respectively. Thus, these scenarios were subjected to sensitivity analysis. Since the CD/BIGCC and VD/CHPSC were not as technically and economically viable, they were not included in the sensitivity analysis. With regards to the sensitivity of the overall ethanol output on profitability (Figure 4), a yield improvement of 10% generally increased the measured IRRs by 1.89%, which improved the IRR of the most optimal ethanol scenarios to $31.29 \pm 0.29\%$. Alternatively, a yield that worsened by 10% reduced the IRRs by 2.02% on average to $27.38 \pm 0.39\%$. The most important observation here is that a reduced yield had only increased the risk of the IRR receding below 25% by a margin of 5.68% at most (for the CD/CHPSC). Thus, the most economically viable ethanol scenarios remained desirable for investment from the private sector when suboptimally performing technology was considered. An increased estimate for the BIGCC installed costs of 10% (Figure 5) had increased the CAPEX for the BIGCC-EE by 10% and the VD/BIGCC by 7.35%. However, since the VD/BIGCC had the highest CAPEX originally, the effect of an increased installed cost was most prominent on this scenario since its IRR was reduced by a further extent than that of the BIGCC-EE. Given a reduction in the IRR of the BIGCC-EE of 1.31%, it is still far more viable than the CHPSC-EE scenario. Regarding the risk around private investment, the VD/BIGCC is still a viable option, as the probability of the IRR receding 25% was only 3.10%.

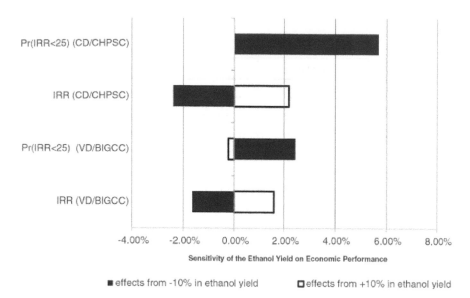

FIGURE 4: Sensitivity of the ethanol yield on the VD/BIGCC and CD/CHPSC scenarios. IRR – Rate of Return; Pr(IRR<25) – probability of IRR falling below 25%.

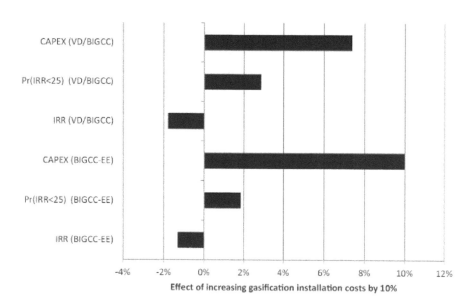

FIGURE 5: Effect of pessimistic installed estimate of gasification costs on the BIGCC-EE and VD/BIGCC scenarios. IRR – Rate of Return; Pr(IRR < 25) – probability of IRR falling below 25%; CAPEX – Capital Expenditure.

13.3 CONCLUSIONS

Ethanol coproduction with electricity generation has been shown to have greater exporting energy efficiency than exclusive electricity generation. This was demonstrated in the context where the status of the electricity generation technologies is advanced, but where the technological status of ethanol production is at a conservative level, considering the modest conversion of sugars to ethanol using the detoxified hydrolysates. If a minimum pricing scenario on the ethanol and electricity products is considered, ethanol coproduction with electricity generation is considerably more profitable than CHPSC power plants, but on par with BIGCC power plants. The advantage of ethanol coproduction would become more pronounced as fermentation technology develops and matures as expected, and it would also become attractive for private investment. If the fermentation of hemicellulose hydrolysates (though successful on a bench-scale) can demonstrate that a pilot-scale conversion of at least 82.5% sugar to ethanol can be attained, it will confirm the technology proposed.

While a high premium on electricity would promote exclusive electricity production when ethanol prices are at a minimum, it is not likely that these premiums on electricity would be attained under the current regulations for renewable electricity in South Africa. The reason is because even at lower premiums, the IRR of the BIGCC-EE was 30%, whereas the prices of electricity for IPPs are regulated to allow for a maximum IRR of 17%.

The study showed that at the current scale, additional capital investments for more energy efficient technology is only justified when it effects a significant improvement in energy efficiency, as shown when BIGCC technology is used instead of the direct combustion in the ethanol coproduction schemes. Furthermore, it was found that when a more energy efficient technology only effected a minor improvement in energy efficiency, such as VD applied in the ethanol coproduction process or biomass dehydration in a combustion process, the economic returns did not justify the capital investment. There would be situations however, where the energy intensity of the process would demand the implementation of such measures, either to reduce process steam demand or improve steam generation to a feasible operating range.

13.4 METHODS

13.4.1 BASIS FOR SIMULATION AND ASSUMPTIONS

All of the scenarios analyzed in the present study assume that the steam demand of the sugar mill itself will be 0.4 tons per ton of cane processed The capital investments required to upgrade existing sugar mills in South Africa to achieve this level of energy efficiency has been estimated at US$ 17.32 million [3,50] (in 2012) for a 300 ton per hour crushing rate. The technical measures included the optimisation of imbibitions rate to reduce the amount of evaporation needed [50]; the conversion from the batch pans to continuous pans and reduce the pan movement water [3,9,50]; using a five-effect evaporator where vapour is bled to the vacuum pans at a lower effect [9]; optimisation of the flashing of condensates for steam recovery [50] and finally; electrifying the turbine drivers [3,4].

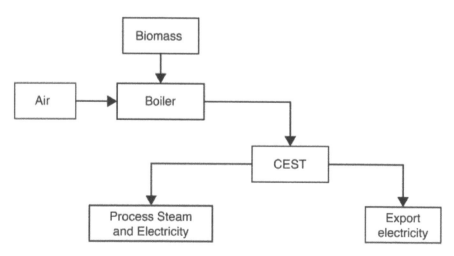

FIGURE 6: Combustion with High Pressure Steam cycles (CHPSC) flow sheet. (CEST – Condensing Extraction Steam.

The cane crushing capacities of mills in South Africa range from 190 to 600 tons per hour [51,52], and thus the representative average of 300 tons per hour was assumed for the present study. Based on a fibre content of 0.14 kg/kg of cane [5], 42 tons per hour of bagasse would be generated from the cane crushing activities for energy generation. Of the total harvesting residues, amounting to 1.167 kg per kg of bagasse [53], 50% of this amount would be collected and co-fed with all the bagasse into an 'energy island' in the mill to generate the energy requirements of the mill (40% steam on cane [41] and 41.64 kWe [4] per ton of cane) and the export energy products. The composition of the residues is shown in Table 3.

13.4.2 PROCESS TECHNOLOGY CONSIDERATIONS

13.4.2.1 EXCLUSIVE ELECTRICITY GENERATION FROM LIGNOCELLULOSIC BIOMASS

Nsaful et al.[41] developed a flow sheet for the conversion of bagasse from a sugarcane milling operation to steam and electricity with a high pressure steam system using combustion (Figure 6). They found that the optimum boiler pressure for efficient electricity generation was 82 bar. Based on the net amount of export electricity, the electrical generation efficiency was 21.5% for 82 bar and 20% for 63 bar. Conversations with experts in the South African sugar industry indicate that the design pressure for boilers which will be used to retrofit the sugar mills is 86 bar. Consequently, a boiler pressure of 86 bar is assumed in this study. Conventionally, biomass would enter the combustor at about 45% moisture, though it could also be dried with flue gas to improve the boiler efficiency, with additional capital charges. The minimum acceptable moisture content for bagasse is 30%, so as to avoid self-ignition and/or a dust explosion [54]. Excess air is provided to the boiler, which has been preheated to 250°C [41] by the stack gas, to improve the overall efficiency. The amount of air in excess is determined to ensure a minimum oxygen content of 6% in the flue gas, as per environmental regulations [55].

The boiler would generate superheated steam at about 515°C [10] and 86 bar that would be expanded in a CEST to generate electricity. The

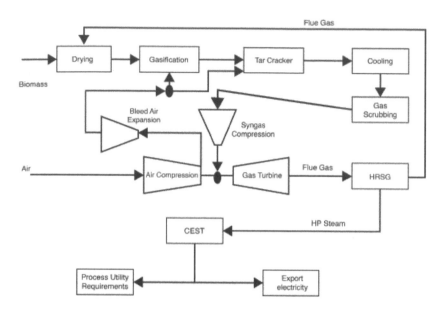

FIGURE 7: Biomass Integrated Gasification and Combined Cycle flow sheet. (CEST – Condensing Extraction Steam Turbine; HRSG – Heat Recovery Steam Generator; HP Steam – High Pressure Steam).

intermediate pressures in the CEST are 13 and 4 bar, and the final vacuum pressure is 0.2 bar [56]. The vacuum steam would then be condensed and returned to the steam cycle. Steam for the mill would be extracted from the CEST at 4 bar. Regarding the performance of the turbines, the isentropic efficiency was 85% while the combined mechanical and electrical efficiencies were set at 96.06% (i.e. 98% for electrical and 98% for mechanical [57]).

The general flow sheet of the BIGCC system (Figure 7) for biomass conversion to electricity was adapted from models developed by the NREL [46,58] and previous reports [59-61]. The combined bagasse and trash would be initially dried to a moisture content of 10 to 15% with exhaust flue gasses [59,60]. The moisture content within the biomass would serve as the gasification control agent [59] since steam injection is generally not considered for gasifier applications in the BIGCC systems. The amount of air added to the circulating fluidized bed (CFB) gasifier in the BIGCC system is to ensure the highest possible calorific value of the syngas without an excess of tars. To initiate tar cracking, the gasifier is required to be operated at 800°C. This is achievable when the ratio of the air supplied to the stoichiometric amount for complete combustion (equivalence ratio) is 0.25 to 0.3 [59]. The syngas would then enter a cyclone to remove particulate matter before entering the CFB tar cracker, where additional air is added to increase the temperature to 920°C [46] for cracking to occur with a dolomite catalyst [46].

After tar cracking, the temperature of the syngas would be reduced to 288°C in order to condense the alkali species for removal with any other particulate matter in a filter bag [46]. The syngas is further cooled to 97°C before it is scrubbed with water to remove the nitrogenous and sulfurous compounds for the protection of downstream equipment, and to prevent nitrous and sulfurous oxide emissions [46,60]. The scrubbing also humidifies the syngas, which assists in the control of the temperature in the gas turbine [62].

The syngas is compressed in a multi-stage compressor (two compressors with an inter-stage cooler in-between) to 20 bar, which is 4 bar above the combustion pressure of the gas turbine [63], in order to allow for the pressure drop across the feed nozzle [61]. At the inter-stage cooler, the syngas was cooled to 97°C at an intermediate pressure of 6 bar, which was determined by the optimization procedure of Polyzakis et al.[63]. The air required for the gas turbine would be compressed in the compressor chamber of the gas turbine, and is fed at a mass ratio of 1:5.14 to the syngas [64]. In order to compensate for the extra volumes of dilution gasses in the syngas that is not found in natural gas, such as CO_2 and nitrogen, air would be bled from the compressor chamber before the air enters the combustion chamber [46,60,64] since the gas turbines are designed for natural gas which does not contain inert gasses. The rate at which air is bled amounts to 13.3% of the air fed to the compression chamber, which is in excess of the air demands of the gasifier and cracker. The bleed air would be expanded through a turbine to atmospheric pressure to improve the net electricity output [60]. Once expanded, the air would be preheated to the gasifier temperature to be used to feed the gasifier and tar cracker. Regarding the efficiency of the gas turbine, the compressor section has a polytrophic efficiency of 0.87% and the gas turbine has an isentropic efficiency of 89.77% [57]. The mechanical and electrical efficiencies are both set at 98% [57].

The combustion exhaust gas leaves the combustion chamber at about 1100°C and is expanded to 1 bar in the turbine section to generate the bulk of the electrical output by driving the generator. The heat of exhaust gasses of the turbine is captured in a HRSG to generate steam at 60 bar for the 'combined' steam cycle [58,60,61]. The superheated steam is generated with water that had been preheated with waste heats from the BIGCC system, as determined by PPA. The superheated steam is then expanded in a CEST to provide the steam demands and additional electricity. The exhaust gas of the HRSG has a temperature of 200°C and is used to dry the biomass that entered the BIGCC system [60].

13.4.2.3 ETHANOL COPRODUCTION FROM HEMICELLULOSE

A general overview of the process flow sheet for the coproduction of ethanol and electricity is presented in Figure 8. Regarding the process step for pretreating the biomass for hemicellulose solubilization, the two methods that have been established for this purpose are dilute acid hydrolysis (DAH) and steam explosion (STEX). As shown in Table 4, the DAH process typically provided a lower yield of hemicellulose sugars than the STEX and only operated effectively at solid contents of 20% and below. STEX had been shown to operate effectively at 50% (Table 5), which implies that the energy requirements for DAH were much higher due to the large volumes of water to be evaporated to concentrate the hydrolysate for efficient downstream processing [47,65]. Thus, STEX was selected as the technology for hemicellulose solubilization. After pretreatment, the pretreated slurry is washed and filtered to recover the solubilized sugars as a filtrate from the solid residue (containing the cellulose and lignin fractions) that is converted to heat and power (via CHPSC or BIGCC). The sugar recovery to the filtrate was conservatively estimated at 80%, even though recoveries as high as 91% have been reported in the literature [21]. Regarding the gasification of residue from steam explosion, it produces a syngas with a higher calorific value than that produced from raw biomass,

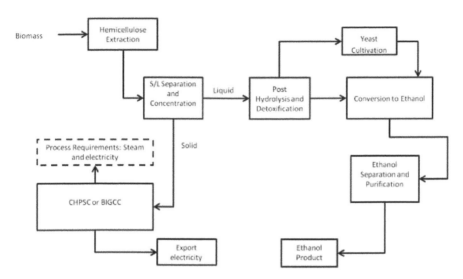

FIGURE 8: Ethanol coproduction flow sheet. (BIGCC - Biomass Integrated Gasification and Combined Cycles system; CHPSC – Combustion with High Pressure Steam Cycle; S/L Separation – Solid Liquid Separation.

but the gasifier must be slightly bigger, to accommodate the lower reactivity of the pretreated biomass [66].

STEX without catalysis requires a secondary hydrolysis of the hydrolysate to convert the oligomers to fermentable monomeric sugars [76] by treatment with sulfuric acid at a level of 0.5% (wt) in solution at 120°C for 20 minutes [76]. This two-step process is advantageous over the catalyzed STEX (and DAH) because catalysis is intended for improving enzymatic digestibility (which is of no significance in this study) rather than the solubilisation of hemicellulose [77]. Furthermore, the yield of solubilisation with no catalysis is less variable, as shown in Table 6 and lastly, the costs of purchasing or producing SO_2 and effective impregnation equipment are avoided.

The theoretical conversion of hemicellulose of the mixed biomass is assumed as 88%, which is the weighted average obtained with bagasse [72,73] and trash [49] (at a feed ratio of 1:0.583) at temperatures ranging from 200 to 210°C with a STEX time of 10 minutes. At such conditions

TABLE 4: Comparison of hemicellulose solubilization techniques (WIS – Water insoluble solids).

Dilute acid hydrolysis	Hemicellulose %	WIS %	Solubilisation %	Yield g/100 g
Aguilar et al. [67]	21	10	92	19.0
Canilha et al. [19]	26	-	62	15.9
Tricket [68]	30	19	71	21.3
Diedericks et al. [48]	24	30	75	18
Lavarack et al. [69]	28	20	72	20.0
Steam Explosion	Hemicellulose %	WIS %	Solubilisation %	Yield g/100 g
Rudolf et al. [26] (SO$_2$ catalysed)	26	50	87	23
Carrasco et al. [70] (SO$_2$ catalysed)	24	50	57	16
Ferreira-Leitao et al. [71] (CO$_2$ catalysed)	23	50	63	14
Laser et al. [72] (uncatalysed)	26	50	91	24
Rocha et al. [73] (uncatalysed)	25	50	82	21
Oliveira et al. [49] (uncatalysed) (trash)	29	15	93	27

TABLE 5: Fermentation parameters.

Parameter	Value and description
Hydrolysate	Alkaline detoxification [24,25]
Total dissolved sugars in raw hydrolysate	200 g/l primarily consisting of hemicelluloses. Glucose originating from 30% solubilisation of the cellulose in the trash [49]
Fermentation Mode	Fed-batch, initially with 60% of the reactor loaded with diluted hydrolysate
Fed-Batch Initial Sugars	92 g/l, as per the hydrolysate in Kurian et al.[24]
Yeast loading	1.5-2 g/l dry weight [25,74]
Conversion to Ethanol	82.5% of pentose, 90% of hexose
Fermentation Time	140 hours [75]

the formation of degradation compounds is significant enough to inhibit the fermentation organisms [72], and thus the neutralization of these compounds by detoxification is necessary. The mode of detoxification would be alkaline treatment [24,25,28] at temperatures below 30°C so that no sugar loss occurs [79]. While detoxification has been criticized as an unnecessary economic hurdle in previous process evaluations, it has been shown experimentally that detoxification improves the yield of fermentation by 20 to 25%, even if inhibitor resistant strains are used [25,80]. In this study, ammonium hydroxide will be used to carry out detoxification [81] in order to avoid the environmental and operational issues associated with over-liming. The yield of ethanol on sugar and other fermentation parameters for the present study are presented in Table 5, which are conservatively assumed based on previous performances in literature [26,27]. To ensure this yield, fed-batch fermentation is adopted, since this mode of fermentation can be 1.5 times more efficient than batch fermentation [82]. With regards to the hexose sugars, which resulted primarily from the cellulose content of the sugarcane trash, the conversion will be assumed as 90% since the yeasts consume these sugars at a much faster rate than the pentose sugars [25].

After fermentation, the beer product is purified in the refining section to produce anhydrous ethanol. The beer is initially flashed at 86°C and 0.83

TABLE 6: Economic parameters.

Parameter	Description
Plant Life [47]	25 years, with 9 operational months per year. The salvaging value is 20%.
Period of Economic Analysis [37]	20 years
Depreciation [37,47]	Straight line to salvaging value.
Tax [78]	South African company tax of 28%
Working Capital [47]	5%
Other	To simplify the analysis, an equity of 100% was assumed [28,47]. It was further assumed that capital will be fully paid after construction.

bar to remove the carbon dioxide (CO_2), and the flash gas enters a knockout drum at 1.1 bar to condense ethanol that has evaporated [47,65,83]. The gas from the knockout drum is then combined with the fermentation vent stream and enters a scrubber to ensure the maximum recovery of ethanol [47,65,83]. The effluents from the scrubber and knockout drum are combined with the beer stream from the initial flash and fed to the beer column, which produces a vapor phlegm [29] that contains ethanol at 40 to 50%. The phlegm is then fed to a rectifier column that produces hydrous ethanol phlegm of 91% [83]. The hydrous ethanol phlegm is dehydrated by molecular sieves to produce ultra-pure ethanol product of 99.7% [47,65,83].

In a variation of the configuration described, the beer column can operate at a vacuum pressure of 0.2 bar, which allows for the heat required by the reboiler of beer column to be supplied by the condenser of the rectifier. Dias et al.[29] had shown through pinch-point heat integration that the steam demand of the refining section of an ethanol distillery can be significantly reduced if such a strategy is employed. This variant would, however, require a vacuum pump to actuate, which implies that the economic impacts of the lower steam demand would need weighing-up with the higher electricity and capital requirement.

13.4.2 SCENARIO DEVELOPMENT AND SIMULATION

13.4.2.1 DEVELOPMENT OF SCENARIOS

For electricity and steam generation, technological variants included biomass integrated gasification with combined cycles (BIGCC) and combustion with high pressure steam cycles (CHPSC). Regarding ethanol purification, the technological variants included conventional distillation (CD) and vacuum distillation (VD). Therefore, from these process technologies, the alternative process scenarios that were modelled as all possible combinations of the process options included BIGCC exclusive electricity generation (BIGCC-EE); CHPSC exclusive electricity generation (CHPSC-EE); CD/BIGCC for coproduction of ethanol and electricity; CD/CHPSC for coproduction of ethanol and electricity; VD/BIGCC for coproduction of ethanol and electricity and VD/CHPSC for coproduction of ethanol and electricity.

13.4.2.2 TECHNICAL SIMULATION

The simulation of all scenarios were completed in Aspen Plus® (Aspen Technology, Inc., Massachusetts, USA) [84] by modelling the flow sheets with the relevant design parameters discussed in the section dealing with process technology considerations. Reactors were simulated as the stoichiometric reactor 'RSTOIC' using conversions evaluated from available experimental data [28,37,47,65]. The gasifier in the BIGCC scenarios was modelled as a combination of an adiabatic combustor and equilibrium reactor that determines that determines the gas composition by minimizing the Gibbs Free Energy 'RGIBBS', since the composition of the syngas is close to the equilibrium values [46,59,85,86]. Regarding thermodynamic properties, the Non-Random Two-Liquid model (NRLT) for electrolyte systems (ELECNRTL) was used whenever there were electrolytes to consider, and the NRLT was used to model the separation systems [29,47].

The utility requirements were determined with PPA, by importing heat duties of streams and flash drums determined by the Aspen Plus®

simulations into the IChemE pinch analysis spreadsheet [87] which were used to calculate the hot and cold utilities according to the methods of Kemp [31] and March [32]. After the utility requirements were taken into account, the net export efficiency (η) was calculated as the combination of the net export of electricity (Enet) and calorific value of the ethanol fuel (F) relative to the calorific input of the biomass (m*HHVbiomass), as shown in Equation 1 (the standard unit used for these energy quantities was MJ/hr).

$$\eta = \frac{E_{net} + F}{m \times HHV_{biomass}} \qquad (1)$$

13.4.3 FINANCIAL RISK ASSESSMENT

13.4.3.1 METHOD AND ASSUMPTIONS

Results from the process models were used in the economic evaluation models for each of the scenarios considered, in accordance with established process evaluation techniques [88]. The information emanating from these models were used to calculate the key economic variable (KEI), namely, IRR [88]. A Monte Carlo simulation was super-imposed on the financial evaluation models in order to create a financial risk assessment whereby the input of certain economic variables into the financial statements would be based on probabilistic distributions that are determined by historical data. This process is then repeated for a large number of iterations and the KEIs generated from the iterations are stored for aggregation in a statistical assessment. The methods followed for these simulations have been described in detail by Richardson et al. [34,35,78] and by the authors of the current study [36,37]. The software used for carrying out the simulation was the Simetar Risk Analysis Software (Simetar, Inc., Texas, USA) [89]. The economic parameters that define a South African context under which the financial risk of all processing scenarios were evaluated are given in Table 6.

13.4.3.2 CAPITAL COST ESTIMATES

The capital costs of generic equipment such as pumps, process drums and turbines were estimated with the Aspen Icarus® (Aspen Technology, Inc., Massachusetts, USA) estimator. For all the major or specialized equipment in the ethanol process model, the costs were based on vendor and literature based quotes, such as those in Humbird et al.[74] for steam explosion, hydrolysis equipment and filter presses; Aden et al.[28] for detoxification equipment, cooling water mains pump and cooling tower system; Bailey [90] for surface condensers; Ridgway [91] for vacuum pump; Al-Riyami et al.[92] for heat exchanger costs and finally, Craig and Mann [46] for flue gas biomass dryer.

The cost of the high pressure boiler systems was provided by experts in the South African sugar industry. However, for the BIGCC it was seen that capital estimates tended to be based on whole plant costs in the literature [16,93]. Many of these estimates could be traced to vendor quotes from the period when the costs of BIGCC systems were still pioneer costs [13,16,93]. Therefore, the most relevant estimate was found in the representative costs in the Report on Combined Heat and Power, of the Environmental Protection Agency of the USA (EPA CHP) [42], which was based on vendor quotes of modern equipment costs.

13.4.3.3 NOMINAL ECONOMIC VARIABLES

In this study, nominal economic variables' refers to specific prices and indices that form the basis of operating costs, incomes and interest-based transactions. These specific values were either determined from the literature or from published databases and are listed in Tables 7 and 8. The variables are treated as 'static' variables (Table 7), which means that the basic prices in year one of the assessment was taken as an average value estimates and inflated with the projected Producers Purchases Index (PPI) to predict the future value. The 'stochastic' variables (Table 8) were used to generate a multivariate empirical function, from which the future val-

TABLE 7: Static nominal economic expenses.

Item	Cost basis**	Value
Maintenance and Repair [78]	US$/litre	0.004
Labour Ethanol plant [78]	US$/litre	0.016
Management and Quality Control [78]	US$/litre	0.005
Real Estate Taxes [78]	US$/litre	0.001
Licenses, Fees and Insurance [78]	US$/litre	0.001
Miscellaneous Expenses [78]	US$/litre	0.005
Total Chemical Cost (per annum)	US$	2, 850, 292
CHPSC Operating Costs [42]	US$/litre	0.007
BIGCC Operating Costs [42]	US$/litre	0.013
Trash Price [7]	US$/dry ton	20.44
Bagasse Price[1]	US$/dry ton	6.3

**All prices are given for the year 2012 in the South African Market
[1]Amortized cost for upgrading a South African mill to 40% steam on cane for liberating bagasse.

ues were iteratively projected for each year in the evaluation in the Monte Carlo simulation.

Regarding the price of second generation ethanol, uncertainty exists because the current pricing of the South African biofuel strategy does not explicitly include second generation fuels [100]. Thus, the price of ethanol was either based on data given for ethanol prices in the USA, or based on data describing Brazilian (BRZ) ethanol prices. Given the uncertainty in the selling price of the export electricity, an upper and lower premium was calculated and applied on the base electricity prices projected from the probability distribution. These premiums were calculated on the minimum and maximum prices for renewable electricity of the South African Renewable Energy bids, which had 2012 based prices of 0.113 and 0.175 US$ per kW/hr respectively [101]. Since biomass based energy is continuous and supplies energy for peak hours, a bonus price of two times the renewable energy price is allowed for four hours per day [101], which thus raises the average renewable prices to 0.139 and 0.216 US$ per kW/

TABLE 8: Database for stochastic variables.

Unit	Electricity [94] US$ per kWhr	Brazilian ethanol [95] US$/litre	USA ethanol [96,97] US$/litre	PPI [98]	Interest rate [99] %
2003	0.032	0.252	0.337	124.80	15.16
2004	0.039	0.254	0.422	127.70	11.31
2005	0.044	0.375	0.463	132.40	10.64
2006	0.048	0.508	0.674	142.60	11.14
2007	0.040	0.467	0.524	158.20	13.08
2008	0.040	0.520	0.587	180.80	15.12
2009	0.045	0.450	0.449	180.70	11.80
2010	0.054	0.612	0.483	191.60	9.91
2011	0.087	0.867	0.683	207.60	9.00
2012	0.106	0.666	0.611	220.50	8.78

hr. Relative to the base electricity price (2012 price) in Table 8, the upper and lower premiums amount to 31% and 103% respectively.

With the two possible renewable electricity premiums and two sets of ethanol price data, there would be four possible pricing strategies. Each strategy is considered as a set of parameters under which separate sets of stochastic evaluations of the process scenarios will be conducted. These four pricing strategies are: 1) minimum electricity premium and ethanol prices based on Brazilian data, 2) minimum electricity premium and ethanol prices based on USA data, 3) maximum electricity premium and ethanol prices based on Brazilian data, and 4) maximum electricity premium and ethanol prices based on USA data.

13.4.4 SENSITIVITY ANALYSIS

The modern capital estimates of BIGCC carry some uncertainty, in that they might be too optimistic. Thus, in order to account for the possibility of a more pessimistic capital estimate, the capital estimates of the BIGCC power plant equipment was increased by 10%, prior to factoring in balance of plant costs (BOP) and project contingencies. Accordingly, a separate stochastic simulation of the BIGCC-EE, CD/BIGCC and VD/BIGCC scenarios was carried out to evaluate the impact of the pessimistic capital estimate.

There is also uncertainty in the yield of ethanol from the hydrolysate, as a very sophisticated organism could yield more ethanol without the need for detoxification, hence negating the associated costs. Furthermore, it is also possible that suboptimal hemicellulose extraction could result in a loss of ethanol. Thus, two sensitivity scenarios of the most profitable co-production scenarios were simulated. They include: (1) where the overall ethanol yield is decreased by 10% to account for the suboptimal hemicellulose extraction, and (2) where the overall ethanol yield is increased by 10% to account for a possibility of a sophisticated fermenting organism which will further discard the costs associated with detoxification. Accordingly, the economic sensitivity to these variations in process parameters was determined using separate stochastic evaluations.

13.5 ABBREVIATIONS

BIGCC: Biomass Integrated Gasification and Combined Cycle; BOP: Balance of Plant; BRZ: Brazil; CD: Conventional Distillation; CEST: Condensing Extraction Steam Turbine; CFB: Circulating Fluidized Bed Reactor; CHPSC: Combustion with High Pressure Steam Cycle; DAH: Dilute Acid Hydrolysis; EE: Exclusive Electricity Production; Enet: Net Export Electricity; F: Ethanol Fuel; HRSG: Heat Recovery Steam Generator; IRR: Internal Rate of Return; KEI: Key Economic Indicator; n: net export efficiency; NPV: Net Present Value; NREL: National Renewable Energy Laboratory; PPA: Pinch Point Analysis; PPI: Producers Purchase Index; SA: South Africa; STEX: Steam Explosion; TCI: Total Capital Investment; TR: Technical Related; USA: United States of America; VD: Vacuum distillation.

REFERENCES

1. Mbohwa C, Fukuda S: Electricity from bagasse in Zimbabwe. Biomass and Bioenergy 2003, 25:197-207.
2. Dias MOS, Cunha MP, Jesus CDF, Rocha GJM, Geraldo J, Pradella C, Rossell CEV, Maciel R, Bonomi A: Second generation ethanol in Brazil: can it compete with electricity production? Bioresour Technol 2011, 102:8964-8971.
3. Peacock S: Process Design for Optimum Energy Efficiency. Glenashley, South Africa: Tongaat Hullet Sugar, Corporate Head Office; 2008.
4. Venkatesh KS, Roy AS: Development and installation of high pressure boilers for co-generation plant in sugar industries. Smart Grid and Renewable Energy 2010, 1:51-53.
5. Ensinas AV, Nebra SA, Lozano MA, Serra LM: Analysis of process steam demand reduction and electricity generation in sugar and ethanol production from sugarcane. Energy Conversion and Management 2007, 48:2978-2987.
6. Botha T, Von Blottnitz H: A comparison of the environmental benefits of bagasse-derived electricity and fuel ethanol on a life-cycle basis. Energy Policy 2006, 34:2654-2661.
7. Dias MOS, Junqueira TL, Cavalett O, Cunha MP, Jesus CDF, Rossell CEV, Maciel R, Bonomi A: Integrated versus stand-alone second generation ethanol production from sugarcane bagasse and trash. Bioresour Technol 2012, 103:152-161.
8. Huang H-J, Ramaswamy S, Al-dajani W, Tschirner U, Cairncross RA: Effect of biomass species and plant size on cellulosic ethanol: a comparative process and economic analysis. Biomass and Bioenergy 2009, 33:234-246.

9. Ogden J, Hochgreb S, Hylton M: Steam economy and cogeneration in cane factories. Intenational Sugar Journal 1990, 92:131-140.

10. Moor B: Modern Sugar Equipment Assisting Cogeneration. In Proceedings of the South African Sugar Technologists' Association (SASTA) Congress. Mount Edgecombe, South Africa: SASTA; 2008:1-13.

11. Macrelli S, Mogensen J, Zacchi G: Techno-economic evaluation of 2nd generation bioethanol production from sugar cane bagasse and leaves integrated with the sugar-based ethanol process. Biotechnol Biofuels 2012, 5:22.

12. Bridgwater AV, Toft AJ, Brammer JG: A techno-economic comparison of power production by biomass fast pyrolysis with gasification and combustion. Renew Sustain Energy Rev 2002, 6:181-248.

13. Caputo AC, Palumbo M, PMP Ã, Scacchia F: Economics of biomass energy utilization in combustion and gasification plants: effects of logistic variables. Biomass and Bioenergy 2005, 28:35-51.

14. Wang L, Weller CL, Jones DD, Hanna MA: Contemporary issues in thermal gasification of biomass and its application to electricity and fuel production. Biomass and Bioenergy 2008, 32:573-581.

15. Searcy E, Flynn P: The Impact of biomass availability and processing cost on optimum size and processing technology selection. Appl Biochem Biotechnol 2009, 154:271-286.

16. Uddin SN, Barreto L: Biomass-fired cogeneration systems with CO 2 capture and storage. Renew Energy 2007, 32:1006-1019.

17. Faaij A: Modern biomass conversion technologies. Mitig Adapt Strat Glob Chang 2006, 11:343-375.

18. Kazi FK, Fortman JA, Anex RP, Hsu DD, Aden A, Dutta A, Kothandaraman G: Techno-economic comparison of process technologies for biochemical ethanol production from corn stover q. Fuel 2010, 89:S20-S28.

19. Canilha L, Santos VT, Rocha GJ, e Silva Almeida J, Giulietti M, Silva S, Felipe MG, Ferraz A, Milagres AMF, Carvalho W: A study on the pretreatment of a sugarcane bagasse sample with dilute sulfuric acid. J Ind Microbiol Biotechnol 2011, 38:1467-75.

20. Purchase BBS, Walford SN, Waugh EJ: An update on progress in the production of ethanol from bagasse. Proceedings of the South African Sugar Technologists' Association - June 1986, 1986(June):33-36.

21. Treasure T, Gonzalez R, Venditti R, Pu Y, Jameel H, Kelley S, Prestemon J: Co-production of electricity and ethanol, process economics of value prior combustion. Energy Conversion and Management 2012, 62:141-153.

22. Gnansounou E, Dauriat A, Wyman CE: Refining sweet sorghum to ethanol and sugar: economic trade-offs in the context of North China. Bioresour Technol 2005, 96:985-1002.

23. Gírio FM, Fonseca C, Carvalheiro F, Duarte LC, Marques S, Bogel-Lukasik R: Hemicelluloses for fuel ethanol: a review. Bioresour Technol 2010, 101:4775-4800.

24. Kurian J, Minu A, Banerji A, Kishore VV: Bioconversion of hemicellulose hydrolysate of sweet sorghum bagasse to ethanol by using Pichia stipitis NCIM 3497 and Debaryomyces hansenii SP. Bioresources 2010, 5:2404-2416.

25. Nigam JN: Ethanol production from wheat straw hemicellulose hydrolysate by Pichia stipitis. J Biotechnol 2001, 87:17-27.
26. Rudolf A, Baudel H, Zacchi G: Simultaneous saccharification and fermentation of steam-pretreated bagasse using Saccharomyces cerevisiae TMB3400 and Pichia stipitis CBS6054. Biotechnol Bioeng 2008, 99:783-790.
27. Mcmillan JDJ: Progress on Advanced Liquid Biofuels in the USA Biofuels. In Presented at International Symposium on Alchohol Fuels (ISAF); 2013. Stellenbosch, South Afrca: ISAF; 2013.
28. Aden A, Ruth M, Ibsen K, Jechura J, Neeves K, Sheehan J, Wallace B, Montague L, Slayton A, Lukas J: Lignocellulosic Biomass to Ethanol Process Design and Economics Utilizing Co-Current Dilute Acid Prehydrolysis and Enzymatic Hydrolysis for Corn Stover Lignocellulosic Biomass to Ethanol Process Design and Economics Utilizing Co-Current Dilute Acid Prehyd. Colorado, USA: National Renewable Energy Laboratory (NREL); 2002.
29. Dias MOS, Modesto M, Ensinas AV, Nebra SA, Maciel R, Rossell CEV: Improving bioethanol production from sugarcane: evaluation of distillation, thermal integration and cogeneration systems. Energy 2011, 36:3691-3703.
30. Grisales R, Cardona CA, Sanchez O, Guterrez L: Heat Integration of Fermentation and Recovery Steps for Fuel Ethanol Production from Lignocellulosic. Costa Verde, Brazil: 4th Mercosur Congress on Process Systems Engineering, EMPROMER; 2005.
31. Kemp I: Pinch Analysis and Process Integration: A User Guide on Process Integration for the Efficient Use of Energy. Butterworth-Heinemann, Burlington, USA: Elsevier; 2007.
32. March L: Introduction to Pinch by Linnhoff March. Northwich, United Kingdom: Linnhoff March LTD; 1998.
33. Sue D-C, Chuang C-C: Engineering design and exergy analyses for combustion gas turbine based power generation system. Energy 2004, 29:1183-1205.
34. Richardson JW, Herbst BK, Outlaw JL, Anderson DP, Klose SL, Gill RC II: Risk Assessment in Economic Feasibility Analysis: The Case of Ethanol Production in Texas. Texas, USA: Agricultural and Food Policy Centre; 2006.
35. Richardson JW, Klose SL, Gray AW: An applied procedure for estimating and simulating multivariate empirical (mve) probability distributions in farm-level risk assessment and policy analysis. Journal of Agricultural and Applied Economics 2000, 2(August):299-315.
36. Amigun B, Petrie D, Görgens J: Economic risk assessment of advanced process technologies for bioethanol production in South Africa: Monte Carlo analysis. Renew Energy 2011, 36:3178-3186.
37. Petersen A: Comparisons of the Technical, Financial Risk and Life Cycle Assessments of Various Processing Options of Sugarcane Bagasse To Biofuels In South Africa. South Africa: Stellenbosch University; 2012.
38. Pellegrini LF, de Junior SO, de Burbano JC: Supercritical steam cycles and biomass integrated gasification combined cycles for sugarcane mills. Energy 2010, 35:1172-1180.
39. South African Petroleum Industry Association (SAPIA): Annual Report. Sandton, South Africa; 2012.

40. Eskom: Grid Access: Scaling up for large scale renewable energy uptake. In Presentation to South African Photovoltaic Industry Association (SAPVIA). Cape Town, South Africa: South African Photovoltaic Industry Association.

41. Nsaful F, Görgens JF, Knoetze JH: Comparison of combustion and pyrolysis for energy generation in a sugarcane mill. Energy Conversion and Management 2013, 74:524-534.

42. EPA: Biomass CHP Catalogue, Part 7: Representative Biomass CHP System Cost and Performance Profiles. Washington DC, USA: Combined Heat and Power Partnership; 2007.

43. Dornburg V, Faaij PC: Efficiency and economy of wood-fired biomass energy systems in relation to scale regarding heat and power generation using combustion and gasification technologies. Biomass and Bioenergy 2001, 21:91-108.

44. National Energy Regulatory of South Africa (NERSA) Consultation Paper Review of Renewable Energy Feed In Tariffs. Pretoria, South Africa: Nersa Office; 2011.

45. Justice S: Private Financing of Renewable Energy – A Guide for Policy Makers. Nairobi, Kenya: United Nations Environment Programme (UNEP); 2009.

46. Craig KR, Mann MK: Cost and Performance Analysis of Biomass-Based Integrated Gasification Combined-Cycle (BIGCC) Power Systems Cost and Performance Analysis of Biomass-Based Integrated Gasification Combined-Cycle (BIGCC) Power Systems. Colorado, USA: National Renewable Energy Laboratory; 1996.

47. Leibbrandt NH: Techno-Economics Study for Sugarcane Bagasse to Liquid Biofuels in South Africa: A Comparison between Biological and Thermochemical Process Routes. Stellenbosch, South Africa: University of Stellenbosch; 2010.

48. Diedericks D, Rensburg EV, Görgens JF: Fractionation of sugarcane bagasse using a combined process of dilute acid and ionic liquid treatments. Appl Biochem Biotechnol 2012, 167:1921-1937.

49. Oliveira FMV, Pinheiro IO, Souto-maior AM, Martin C, Gonçalves AR, Rocha GJM: Industrial-scale steam explosion pretreatment of sugarcane straw for enzymatic hydrolysis of cellulose for production of second generation ethanol and value-added products. Bioresour Technol 2013, 130:168-173.

50. Wienese A, Purchase B, Wienese A, Purchase BS: Renewable Energy: An Opportunity for The South African Sugar Industry? In Proceedings of the South African Sugar Technologists' Association (SASTA), Volume 78. Mount Edgecombe, South Africa: SASTA; 2004:39-54.

51. Van der Westhuizen WA: A Techno-Economic Evaluation of Integrating First and Second Generation Bioethanol Production from Sugarcane In Sub-Saharan Africa. Stellenbosch, South Africa: Stellenbosch University; 2012.

52. South African Sugar Mills http://www.huletts.co.za/ops/south_africa/mills/felixton.asp

53. Seabra JEA, Tao L, Chum HL, Macedo IC: A techno-economic evaluation of the effects of centralized cellulosic ethanol and co-products refinery options with sugarcane mill clustering. Biomass and Bioenergy 2010, 34:1065-1078.

54. Mendes AGP: Dust explosions. In Proceedings of The South African Sugar Technologists' Association (SASTA) Congress: Volume 74, 1999. Mount Edgecombe, South Africa: SASTA; 1999:282-288.

55. Department of Environmental Affairs and Tourism: AQA Implementation: Listed Activities and Minimum Emission Standards. Draft schedule for section 21 Air

Quality Act, Rev;: Department of Environmental Affairs and Tourism. Pretoria, South Africa; 2008.

56. Nsaful F: Process Modelling Of Sugar Mill Biomass to Energy Conversion Processes and Energy Integration of Pyrolysis. Stellenbosch, South Africa: Stellenbosch University; 2012.

57. Kreutz TG, Larson ED, Liu G, Williams RH: Fischer-Tropsch Fuels from Coal and Biomass. In 25th Annual International Pittsburgh Coal Conference: August 2008. Pittsburgh, USA; 2008.

58. Mann MK, Spath SL: Life Cycle Assessment of a Biomass Gasification Combined-Cycle System. Colorado, USA: National Renewable Energy Laboratory; 1997.

59. Li X, Grace JR, Lim CJ, Watkinson AP, Chen HP, Kim JR: Biomass gasification in a circulating Fluidized bed. Biomass and Bioenergy 2004, 26:171-193.

60. Rodrigues M, Walter A, Faaij A: Performance evaluation of atmospheric biomass integrated gasifier combined cycle systems under different strategies for the use of low calorific gases. Energy Conversion and Management 2007, 48:1289-1301.

61. Fagbenle RL, Oguaka ABC, Olakoyejo OT: A thermodynamic analysis of a biogas-fired integrated gasification steam injected gas turbine (BIG/STIG) plant. Applied Thermal Engineering 2007, 27:2220-2225.

62. Lugo-Leyte R, Zamora-Mata JM, Toledo-Vela'zquez M, Salazar-pereyra M, Torres-aldaco A: Methodology to determine the appropriate amount of excess air for the operation of a gas turbine in a wet environment. Energy 2010, 35:550-555.

63. Polyzakis AL, Koroneos C, Xydis G: Optimum gas turbine cycle for combined cycle power plant. Energy Conversion and Management 2008, 49:551-563.

64. Ree RV, Oudhuis AB, Faaij A, Curvers APW: Modelling of a Biomass Integrated Gasifier Combined Cycle (BIG/CC) System with the Flow sheet Simulation Programme Aspen Plus. Petten, Netherlands: Energy research Centre of the Netherlands (ECN); 1995.

65. Leibbrandt NH, Knoetze JH, Gorgens JF: Comparing biological and thermochemical processing of sugarcane bagasse: an energy balance perspective. Biomass and Bioenergy 2011, 35:2117-2126.

66. Gunarathne DS, Chmielewski JK, Yang W: High Temperature Air/Steam Gasification of Steam Exploded Biomass. Livomo, Italy: International Flame Research Foundation; 2013:1-14.

67. Aguilar R, Ramırez J, Garrote G, Vazquez M: Kinetic study of the acid hydrolysis of sugar cane bagasse. J Food Eng 2002, 55:309-318.

68. Tricket R: Utilisation of Bagasse for the Production of C5 and C6 Sugars. Durban, South Africa: University of Natal; 1982.

69. Lavarack BP, Gri GJ, Rodman D: The acid hydrolysis of sugarcane bagasse hemi-cellulose to produce xylose, arabinose, glucose and other products. Biomass and Bioenergy 2002, 23:367-380.

70. Carrasco C, Baudel HM, Sendelius J, Modig T, Galbe M, Zacchi G, Liden G: SO2 catalyzed steam pretreatment and fermentation of enzymatically hydrolyzed sugarcane bagasse. Enzyme Microb Technol 2010, 46:64-73.

71. Ferreira-leitão V, Perrone CC, Rodrigues J, Paula A, Franke M, Macrelli S, Zacchi G: An approach to the utilisation of CO 2 as impregnating agent in steam pretreat-

ment of sugar cane bagasse and leaves for ethanol production. Biotechnol Biofuels 2010, 3:7.

72. Laser M, Schulman D, Allen SG, Lichwa J, Antal MJ, Lynd LR: A comparison of liquid hot water and steam pretreatments of sugar cane bagasse for bioconversion to ethanol. Bioresour Technol 2002, 81:33-44.

73. Rocha GJM, Martín C, Vinícius FN, Gómez EO, Gonçalves AR: Mass balance of pilot-scale pretreatment of sugarcane bagasse by steam explosion followed by alkaline delignification. Bioresour Technol 2012, 111:447-452.

74. Humbird D, Davis R, Tao L, Kinchin C, Hsu D, Aden A, Schoen P, Lukas J, Olthof B, Worley MD: Sexton and DD: Process Design and Economics for Biochemical Conversion of Lignocellulosic Biomass to Ethanol Process Design and Economics for Biochemical Conversion of Lignocellulosic Biomass to Ethanol. Colorado, USA: National Renewable Energy Laboratory; 2011.

75. Preez JC, Driessel BV, Prior BA: Applied microbiology biotechnology ethanol tolerance of Pichia stipitis and Candida shehatae strains in fed-batch cultures at controlled low dissolved oxygen levels. Appl Microbiol Biotechnol 1989, 30:53-58.

76. Shevchenko SM, Chang K, Robinson J, Saddler JN: Optimization of monosaccharide recovery by post-hydrolysis of the water-soluble hemicellulose component after steam explosion of softwood chips. Bioresour Technol 2000, 72:207-211.

77. Martin C, Galbe M, Nilvebrant N, Jonsson L: Comparison of the fermentability of enzymatic hydrolyzates of sugarcane bagasse pretreated by steam explosion. Appl Environ Microbiol 2002, 98–100:699-716.

78. Richardson JW, Lemmer WJ, Outlaw JL: Bioethanol production from wheat in the winter rainfall region of South Africa: a quantitative risk analysis. International Food and Agribusiness Management Review 2007, 10:181-204.

79. Nilverbrant N, Persson P, Reimann A, De Sousa F, Gorton L, Jonsson J: Limits for alkaline detoxification of dilute-acid lignocellulose hydrolysates. Appl Biochem Biotechnol 2003, 105–108:615-628.

80. Nigam JN: Ethanol production from hardwood spent sulfite liquor using an adapted strain of Pichia stipitis. J Biotechnol 2001, 26:145-150.

81. Alriksson B, Horvath I, Sjode A, Nilverbrant N, Jonsson L: Ammonium hydroxide detoxification of spruce acid hydrolysates. Appl Biochem Biotechnol 2005, 121–124:911-922.

82. Tantirungkij M, Izuishi T, Seki T, Yoshida T: Applied microbiology biotechnoiogy fed-batch fermentation of xylose by a fast-growing mutant of xylose-assimilating recombinant Saccharomyces cerevisiae. Appl Microbiol Biotechnol 1994, 8:8-12.

83. Mcaloon A, Taylor F, Yee W, Regional E, Ibsen K, Wooley R, Biotechnology N: Determining the Cost of Producing Ethanol from Corn Starch and Lignocellulosic Feedstocks Determining the Cost of Producing Ethanol from Corn Starch and Lignocellulosic. Colorado, USA: National Renewable Energy Laboratory; 2000.

84. Aspen Technology Inc (Manufacturer): AspenPlus®, AspenIcarus®. 2008. http://www.aspentech.com/products/aspen-plus.aspx

85. Aboyade A: Co-Gasification of Coal and Biomass: Impact on Condensate and Syngas Production. Stellenbosch, South Africa: University of Stellenbosch; 2012.

86. Chen C, Y-qi J, J-hua Y, Chi Y: Simulation of municipal solid waste gasification for syngas production in fixed bed reactors. J Zhenjiang Univ Sci B 2010, 11:619-628.

87. IChemE: Pinch Point Analysis Spreadsheet. Rugby, United Kingdom: Institute of Chemical Engineers; 2006.
88. Peters and Timmerhaus: Analysis of cost estimation profitability, alternative investments and replacements optimum design and design strategy. In Plant Design and Economics for Chemical Engineers. 5th edition. Columbus, USA: McGraw Hill; 1997.
89. Richardson JW, Schumann K, Feldman P: Simeter - Simulation & Econometrics To Analyze Risk. Texas USA: Simetar, Inc; 2008.
90. Bailey D: Issues Analysis of Retrofitting Once-Through Cooled Plants with Closed-Cycle Cooling. Electric Power Research Institute: USA; 2007.
91. Ridgway S: Projected Capital Costs of a Mist Lift OTEC Power Plant. New York, USA: The American Society of Mechanical Engineers; 1984.
92. Al-Riyami BA, Klimes J, Perry S: Heat integration retrofit analysis of a heat exchanger network of a fluid catalytic cracking plant. Applied Thermal Engineering 2001, 21:1449-1487.
93. Rhodes JS, Keith DW: Engineering economic analysis of biomass IGCC with carbon capture and storage. Biomass and Bioenergy 2005, 29:440-450.
94. Eskom Tariffs and Charges http://www.eskom.co.za/CustomerCare/TariffsAnd-Charges/Pages/Tariff_History.aspx
95. CEPEA – Centro de Estudos Avancados em Economia Aplicada http://www.cepea.esalq.usp.br/xls/SaaofmensalUS.xls
96. State of Nebraska Official Website http://digitalcommons.unl.edu/cgi/viewcontent.cgi?article=1033&context=imsediss
97. Center for Agricultural and Rural Development, Historical Ethanol Prices http://www.card.iastate.edu/research/bio/tools/download_eth_csv.aspx
98. Producer Purchases Index http://beta2.statssa.gov.za/timeseriesdata/Excell/P0142.1%20PPI%20to%202012%20-%20discontinued.zip
99. South African Reserve Bank: Historical Exchange and Interest Rates http://www.res-bank.co.za/Research/Rates/Pages/SelectedHistoricalExchangeAndInterestRates.aspx
100. Department of Minerals and Energy: Biofuels Industrial Strategy of the Republic of South Africa. Pretoria, South Africa; 2007.
101. Eberhard A: Grid-connected renewable energy in South Africa. In Presentation at the International Finance Corporation (IFC), Washington 2013 . Washington, USA: IFC; 2013.

AUTHOR NOTES

CHAPTER 1

Acknowledgments

Financial support for the various studies reported in this chapter was provided by UNEP GRID-Arendal, Kakira Sugar Limited and Belgian Technical Cooperation (BTC). The National Environment Management Authority and the National Agricultural Research Organization in Uganda gave the facilities and technical support that enabled the accomplishment of studies reported here.

CHAPTER 2

Competing Interests

The authors declare that they have no competing interests.

Authors' Contributions

MAL planned the conceptual process, performed all the experiments and was responsible for results analysis and manuscript draft. LDG helped with the conceptual process, monosaccharide analysis, results discussion and manuscript draft. CGSK helped with immunolabeling assays and analysis and manuscript draft. RS was responsible for the robotic platforms operation during the saccharification assays and helped on the results analysis. CAR conducted the scanning electron microscopy experiments and analysis. CAL contributed to the eucalyptus barks preparation, analysis and evaluation of results. MAC prepared grass samples and made them available for this study. ERA and ODB performed the NMR experiments and analysis, and contributed to manuscript draft. IP and SMM coordinated the overall study, and contributed to results analysis and writing up the paper. All the authors approved the final manuscript.

Acknowledgments

The authors are grateful to FAPESP and CNPq for the financial support for this work via grants #2010/11135-6, 2009/18354-8, 2010/08370-3, 2008/56255-9 and 2010/52362-5 (FAPESP); and #159341/2011-6, 482166/2010-0 and 490022/2009-0 (CNPq), Projeto INCT do Bioetanol (CNPq/FAPESP). We are grateful to USP for the financial support via NAP Centro de Instrumentação para Estudos Avançados de Materiais Nanoestruturados e Biossistemas and NAP de Bioenergia e Sustentabilidade, and European Community's Seventh Framework Programme SUNLIBB (FP7/2007-2013) under the grant agreement #251132. We are also grateful to the technician Valeria Gazda for her help with furfural and 5-hydroxymethyl-furfural chromatographic analysis and Paul Knox (Plant Probe) for antibodies supply. The electron microscopy work was performed at LME/LNNano/CNPEM, Campinas, SP, Brazil.

CHAPTER 4

Acknowledgment

We thank our professor Fabricia Paula de Faria, who led us in this research, professor Américo José dos Santos Reis from School of Agronomy and Food Engineering, who gave the sugarcane bagasse, and CNPq, which invested in this project.

CHAPTER 5

Acknowledgments

Grant from CESS-CPPRI is sincerely acknowledged. Authors are also thoroughly grateful to Dr. MO Garg, Director, CSIR-IIP and Dr. RM Mathur, Director, CPPRI for providing necessary facilities and constant support.

CHAPTER 6

Competing Interests

The authors declare that they have no competing interests

Author Contributions
AKC planned and performed the biomass pretreatment, enzymatic hydrolysis, ethanol fermentation, as well as the analysis of the results and manuscript writing. AKC also coordinated the overall study. FAFA assisted in biomass characterization, fermentation experiments and helped the manuscript drafting. VA, MJVB and LNR jointly carried out the Raman Spectroscopy, FTIR and FT-NIR analysis and written related text in the manuscript. OVS analyzed all the results and reviewed the manuscript draft. CAR and FCP provided the yeast strains and fermentation methodology. Both analyzed the fermentation results and contributed to the drafting of the text related to fermentation. SSS coordinated the overall study, analysis of results and finalizing the manuscript. All authors suggested modifications to the draft and approved the final manuscript.

Acknowledgment
Authors are grateful to FAPESP (Process n° 2008/57926-4 and 2010/11258-0) for the financial support and CNPq. FAFA gratefully acknowledges CAPES. CAR, VA, MJVB and LNR acknowledge to FAPEMIG, CNPQ and CAPES for the financial support. We are also thankful to Ms. Juliana RG Reis for her technical assistance. Authors also would like to thank Dr. Durval Rodrigues Jr., Dr. Paulo Suzuki from Engineering School of Lorena and Dr. Rogerio Hein, UNESP, Guaratingueta for SEM, XRD and AFM analysis respectively.

CHAPTER 7

Acknowledgment and Funding
The authors are grateful for the support of the U.S. Department of Energy for funding through a Research Grant. No. DE-FC26-08NT01922. Dr. Suhardi received Fulbright Scholarship from the US State Department to conduct this research in Dr. Boopathy's Laboratory.

Competing Interests
The authors declare that they have no competing interests.

CHAPTER 8

Competing Interests
The authors declare that they have no competing interests.

Author Contributions
BB carried out the isolation, characterization and classification of the fungi described in this work and wrote part of the manuscript. FAGT designed all the primers and supplied the equipment for fungi characterization. LMPM supervised the development of the work, analysed and discussed all the results, wrote the manuscript and supplied all the reagents and materials for this work. All authors read and approved the final manuscript.

Acknowledgments
This work was supported by Fundação de Apoio a Pesquisa (FAP/DF) and Conselho Nacional de Desenvolvimento Científico e Tecnológico (CNPq).

CHAPTER 9

Acknowledgment
The authors would like to express gratitude to the Strategic Research Council for support to the Biorefinery project which has funded this work. We further thank Jens Iversen for technical input for the fermentation.

CHAPTER 10

Acknowledgment
The authors would like to thank the FAPESP BIOEN Program and the Brazilian National Council for Scientific and Technological Development (CNPq) for the financial support. The authors would also like to thank Mr. Alonso Constante Escobar (Tino), Mr. Carlos Henrique Manfredi and Mr. Marcelo Nishida for all technical information about 1G ethanol production and electric energy selling auctions.

Authors' Contributions
FF did the process simulations in EMSO, the 2G sector economic analysis and the sensitivity tests. RT provided the economic analysis tools and did

the economic analysis in the base case, for the 1G sector. FP provided the basic configuration of the 1G plant and helped with the economic analysis. RLG helped defining the technological options for the 2G process configuration (pretreatment, hexoses and pentose fermentation). CC, AC and RG worked in process integration (1G + 2G + EE). RG coordinated the work and supervised all other tasks. All authors read and approved the final manuscript.

Competing Interests
The authors declare that they have no competing interests.

CHAPTER 11

Acknowledgment
We are grateful to the FAPESP for providing the financial support under the thematic project -2008/57926-4 and 2010/08066-2.

Competing Interests
The authors declare that they have no competing interests.

Authors' Contributions
ISO carried out the experimental work. ISO, AKC and MBS designed the study and were involved in all discussions, interpretation of data and writing the manuscript. SSS coordinated the overall study, analysis of results and finalizing the manuscript. All authors suggested modifications to the draft and approved the final manuscript.

CHAPTER 13

Acknowledgment
The authors would like to thank Chair of Energy Research at Stellenbosch University, funded by the Department of Science and Technology, for providing financial assistance.

Competing Interests
The authors declare that they have no competing interests.

Authors' Contributions

AMP was the primary investigator into the work. He performed all simulations of the various processes and economic evaluations. He also completed the interpretation of results and was the primary author of the written text. MCA was the internal reviser who checked the paper critically for inconsistencies in the text and structure. He also gave assistance in the structuring of the paper. JFG is the study leader of the research group, reviewed the paper internally and approved the submission from our department. He is also the corresponding author. All authors agree to be accountable for all aspects of the work and will ensure that questions relating to accuracy and integrity be appropriately investigated and resolved. All authors read and approved the final manuscript.

INDEX

9 781774 635506